The Scopes of Work Program

Procedures
and
Standards
to
Increase Quality

Linda Haas Davenport

BuilderBooks™
National Association of Home Builders
1201 15th Street, NW
Washington, DC 20005-2800
(800) 223-2665
www.builderbooks.com

The Scopes of Work Program: Procedures and Standards to Increase Quality
ISBN 0-86718-514-7

Printed in the United States of America

Cataloging-in-Publication Data available from the Library of Congress

For further information please contact:

BuilderBooks™
National Association of Home Builders
1201 15th Street, NW
Washington, DC 20005-2800
(800) 223-2665
www.builderbooks.com

10/01 B. Minich/Data Reproductions 2000
11/02 Data Reproductions 2000

Disclaimers

About the Author

Linda Haas Davenport began her consulting work for the construction, oil, and gas industries. For the past twelve years, Linda was the operations officer of a mid-sized production builder in Atlanta, GA. She recently returned to her hometown of Tulsa, OK to set up a consulting service for the home building industry.

Acknowledgments

The catalyst that caused me to put the scopes of work program into book form was a conversation I had one day with Bob Merz. Bob had returned from a conference where he had been involved in a roundtable discussion with builders. These builders were complaining about warranty and standardization problems, and the damage being done by inept field people and trades before they were spotted and replaced. I told Bob that these builders needed to implement the scopes of work program, which our company had been using for several years to bring all of these issues under control. Bob was unaware of the scopes of work program. He had only seen, and been impressed with, our company's finished product. After I explained the program, Bob encouraged me to write a book and make the scopes of work program available to all builders.

Once I agreed to write the book, Bob generously gave of his time and technical expertise to review each individual scope of work and inspection report. This was no small chore I can assure you. During the writing and production of this book, he offered suggestions and advice that proved to be invaluable. I cannot thank him enough for this support and help.

I would also like to express my most heartfelt thanks to John Goucher of BuilderBooks. As a new author, I do not know if I could have gotten through the production of this book without his help, advice, and support.

This book is produced under the general direction of Thomas M. Downs, NAHB Executive Vice President and CEO, in association with NAHB staff members Adrienne Ash, Assistant Staff Vice President, Knowledge Management; Charlotte McKamy, Publisher, Home Builder Press; John Goucher, Acquisitions Editor; David Rhodes, Art Director; Barbara Minich, Editor, and Toral Patel, Assistant Editor.

Contents

Foreword

Home builders today are in the business of fulfilling expectations, or at best, managing the expectations of those that purchase their goods and services. Over the past 25 years, I have seen builders make great efforts to develop systems that address the various facets of their industry. Many have refined their standard contract forms to address the multitude of potential questions that are posed in negotiations, and most builders have adopted an expressed warranty in order to better define for their customers what course of action will be taken if a situation should arise. In my opinion, things like this have raised the bar of professionalism for the industry as a whole. I believe that, with more refinement, the builders and customers will continue to benefit from this growth.

With home builders' increased concentration on reaching the buying public's needs, I have seen a growing emphasis placed on systems to deliver these goods and services. The need for certain assurances of on-time delivery without sacrificing quality have become critical, particularly with the dependence of the builder or manager on their trade contractors to perform their work in a skillful manner.

As a veteran of this industry, Linda Davenport has devised a program that brings the accountability of the trade contractors and the field full circle by requiring a commitment, up front, on the conditions of performance for various segments of residential construction. She has integrated the prescriptive elements of the building code requirements with the performance-based *Residential Construction Performance Guidelines* to define a specific scope of work for each phase of the construction process, which is the next logical step to managing the consistent delivery of the end product. By presenting the program in such a manner, she has set the challenge from within to ensure that each segment of work is completed correctly before moving forward. In a business where we place a premium on "doing the work right the first time," it is refreshing to see that someone like Linda has taken the time to develop a measure of what is ideal and how to deal with the situation if the work falls short of that mark.

My familiarity with Linda's scopes of work program stems mostly from my association with the accepted standards of the industry and her implementation of those in the specific scopes of work. Linda's program adds a dimension to the traditional requirements of "all work shall be done in a workmanlike manner and pursuant to the applicable building codes." The program does this by clearly defining the next level of what is expected of each and every trade contractor in their specific language from a performance-based standard that is easily understood and generally accepted as the benchmark of the industry.

Satisfaction of warranty requests does little to raise your image with your customer base. Linda's approach of eliminating, or greatly reducing, the need for warranty repairs ultimately increases your bottom line. Compliance with guidelines and standards during the construction process can only yield a better end product without the additional cost of corrective work. In order to implement the program, you will need to commit to all of its phases. If you do, you will maximize the benefits of this program. I have actually witnessed the success of this program, and this internal check ultimately leads to reduced warranty requests, a better bottom line, and a more satisfied customer.

Bob Merz
Construction Arbitration Associates

Introduction

Get together a group of people who are involved in the home building industry and sooner or later, the topic of conversation almost always gets around to the telling of construction horror stories.

I have a story I love, although it's not really a horror story. I call it the case of the "Hidden Linen Closet." The framers had missed framing the linen closet and, of course, the drywallers had closed in the closet without a second thought. The mistake was discovered when the homebuyer called from the closing wanting to know where we had "hidden" her linen closet, and the secret to getting into it.

On a scale of 1 to 10, the house that "smells" must rank as the number one nightmare for a builder (although it often becomes a most entertaining story). I'm no different from any other builder, and my horror story began with a phone call from a real estate agent on site complaining about one of our finished houses that "smelled." I would guess that every construction person (and most of the office staff) had visited the house at one time or another trying to determine what the odor was and where it was coming from. We tried all the normal things. We had the house cleaned, shampooed the carpet, set off odor bombs, and even called in a house deodorizer expert. Nothing worked. The smell kept coming back.

Eventually, we traced the smell to the master suite. We looked everywhere. We pulled up the carpet, took off soffit, tore out the bathroom linen closet, and did everything else anyone could think of. We still couldn't find the source of the smell. Finally, I asked the HVAC contractor to unhook the duct work to check for any problem with or in the ducts. Before the contractor could unhook the first duct, he disturbed the attic insulation and the smell became overpowering. I had the attic insulation over the master suite removed. Yes, that was the source of the smell. Apparently, before the insulation was chopped and blown, a family of small rodents had made a home in the insulation roll and ended up being added to the insulation. Not only was new insulation required, but we had to remove all the drywall over the master suite and replace it. I would prefer not to think about the time and money wasted on this particular house, or the site superintendent who didn't bother to inspect the attic insulation when it was installed.

My personal favorite story is a mixture of funny and horror. I refer to it as, "Why can't I get my furniture up the stairs?" A homeowner had just closed on his home, and was in the process of moving in when I received a phone call from him. He was extremely upset because his movers could not get his furniture up the stairs. He kept telling me the stairs were too small. "Stairs too small? Nonsense," I thought. "Stairs are never too small." There is no need to discuss the next couple of hours spent talking to field personnel and checking plans. Finally, I went out to the home and guess what? The homeowner was correct. The stairs were 3 inches too narrow and his furniture wouldn't fit. The solution to the problem was to remove a double window in the bonus room upstairs, rent equipment, and hoist the man's furniture to the second floor and then through the window opening. (I have no idea what the man will do when he wants to move out of the house.) The problem began with a footing that was not checked before it was dug and poured. This resulted in a slab that was poured 3 inches too narrow, which also was not checked. The problem continued when the framing crew framed the exterior rooms without verifying all dimensions. When the framers reached the interior stairs and did not have enough room for the pre-built stairs they decided to stick frame them.

I'm sure you have your own horror stories to tell. As entertaining as these stories are when we tell them later, they almost never are entertaining while they are happening. Many builders and field personnel don't stop to think that these stories are the result of what I call "built-in warranty."

Built-in warranty is caused by sloppy work and careless mistakes during the construction process that result in warranty problems. It matters little whether the builder is a small company constructing only a few houses a year, or whether an industry giant building thousands of houses nationwide. The problem of built-in warranty plagues builders of every size.

The scopes of work program is a tool that can help builders eliminate built-in warranty. The scopes of work provides builders with guidelines for their staff and trade contractors. These guidelines describe the quality of work and the standards the company demands for each step of the construction process. The scopes of work, however, go beyond just telling people what is expected of them. Using the inspection reports the actual job is walked, checked, and verified against the company's standards by the site superintendent and trade contractor together. Any sloppy work or careless mistakes must be corrected before the next phase of construction is allowed to begin. This program not only eliminates built-in warranty, it insures standardization from house to house and accountability from field staff and trade contractors. (And, it also eliminates those entertaining horror stories.)

A builder who undertakes the development of this program from scratch will spend a lot of time and money, both of which are in short supply for any builder. This book has been written to help builders implement the program as quickly and economically as possible.

In the step-by-step instructions found in the following chapters, I used a mid-sized production builder's business as my example. I realize that not every builder has reached this size and will not have some of the departments I talk about, nor will they already have in place some of the things to which I refer. However, the nice thing about the scopes of work program is that it is scalable. A builder can begin on a small scale using just the scopes of work and inspection reports. Then, as his or her company grows the program can grow with it.

1

What Is the
Scopes of Work Program?

Many years ago, I was involved in developing a method for a builder to standardize his houses, gain accountability from his staff and field personnel, and reduce warranty work. The builder had grown past the point where he could personally supervise construction by visiting each jobsite every day, much less each house. He also had accumulated people, both in the office and in the field, and was struggling with the problems that always result from having a staff.

The program took many months to develop. Then it took many more months to work out the kinks. When it was finally in day-to-day operation, however, the results surprised all of us—with the biggest surprise being an increase in the bottom line.

Over the years, as a consultant, I have helped to implement the program for many builders. For the past several years, I have been responsible for the daily operation of the scopes of work program for a production builder. I speak from first-hand experience when I tell you that the scopes of work program brings:

- Standardization of house construction

- Accountability from field personnel and trade contractors

- Control of costs

- Increase in sales

- Reduction of warranty

- Increase to the bottom line

The scopes of work program is a combination of tools and procedures. It is not just scopes of work or just inspection reports. It is a flow of responsibility that ensures standardization, quality, and correctness in every phase of construction. It begins with determining what the company will require for each trade contractor who is involved in building a house, and then follows those requirements all the way through the building process. Along the way, it puts into place accountability from field personnel and trade contractors and standardization of each step of construction. It also puts into place accountability from the company to the trade contractor. Implementation of the program requires accountability from field personnel and trade contractors and standardizes each step of construction.

The program makes the company accountable to the trade contractor by telling each trade contractor what is expected of him or her before work begins. In addition, trade contractors know that if their work is performed correctly, they will not be called back to the same job time and time again. This, in turn, saves trade contractors money and makes them more willing to quote a good price and continue to work for the company—a major plus for any builder.

Using scopes of work documents

The scopes of work program is simple. Basically, it is a package of information and a set of day-to-day operating procedures. It also is a contract-agreement between the trade contractor and the company. The package of information contains a set of terms and conditions, job requirements and a warranty guarantee within the scopes of work, and pre- and post-inspection reports.

1. The set of **Terms and Conditions** spells out what the company expects of every trade contractor and what the trade contractor can expect from the company. The terms and conditions are in two sections. The first section covers items such as insurance, haz-com programs, safety, and drug-free workplace expectations. The general Terms and Conditions include many of the same items in more detail, plus information on such items as warranties, walk-throughs, quality expectations, and other items the company wants to bring to the trade contractor's attention.

2. Each step in the construction process has a detailed list of **Job Requirements** or **Scopes of Work**. Each list includes general information about the construction phase, specific requirements for the phase, and any additional items required by the company. The scopes of work are extremely detailed and leave little room for poor-quality performance.

3. Checklists called **Pre- and Post-Work Inspection Reports** are used to verify the quality and completeness of the job. The inspection reports require that the site superintendent, or other qualified person, and the trade contractor walk, check, and verify the job together before work begins and again after the job is completed. The inspection reports are used to track the quality and completeness of the job. Payment is not made until the job is 100-percent complete and is of the quality required by the company. This particular section also allows the company quickly spot inept field personnel and trade contractors.

4. A **Warranty Agreement** between the trade contractor and the company requires a guarantee from the trade contractor that they will perform all warranty on their own work. Should the trade contractor refuse to perform his or her warranty work, the agreement allows the company to either backcharge the cost for another trade contractor to perform the work or allows the company to file a claim against the trade contractor's insurance.

Overview of day-to-day operation

The scopes of work program works like this in day-to-day operation. I'll discuss each of these procedures in more detail in later chapters.

1. A new trade contractor comes in to sign up. They are given: a set of terms and conditions, a scope of work for their particular phase of construction, and the inspection reports for that phase. The trade contractors are instructed to read the document and must agree to abide by the requirements. The paperwork must be signed before the trade contractor will be allowed to work for the company.

2. The site superintendent is ready for the trade contractor to come to the job site. Before the trade contractor begins work, the trade contractor and the site superintendent go to the job area and use the pre-work inspection report to verify that the job area is ready for the trade contractor to begin work.

3. The trade contractor then completes the job.

4. The trade contractor and the site superintendent walk the job and use the post-work inspection report to verify that every item found on the inspection report is in fact completed to the company's standards. If the job is completed correctly, both the trade contractor and the site superintendent sign-off on the job. If the job has not been completed correctly, the trade contractor must fix his or her mistakes and again go through the post-work inspection checklist. The process continues until the work meets the company's standards and is 100-percent complete.

5. When the trade contractor turns in the invoice for the job, the pre and post work inspection reports and the invoice are attached to the office copy of the purchase order.

6. Invoices are turned in to the office for payment and the accounts payable staff checks the invoice and the inspection reports for completeness. If they are completed correctly, the invoice is approved for payment.

2

That's Simple Enough, But What Does It Accomplish?

As I said in Chapter 1, the program itself is simple, but it accomplishes a lot of things that are not so simple. Let's look at a few things the program will accomplish for you.

Standardize the construction of houses

First and foremost, the program is a tool that will standardize each step of the construction process. If every person who works on a house must do their job the same way every time, there is no room for "free spirits" to do it their way. You have eliminated the entrepreneurial tendencies of many trade contractors and field superintendents. You also have a yardstick by which to judge the quality of every job at every phase of construction. You have eliminated the learning curve for new trade contractors to do things your way and taken the guesswork out of the process.

Evaluate employees

When used correctly, the pre- and post-work inspection reports become a means to quickly judge field personnel and their competence. The same is true of trade contractors. You don't have to wait until the warranty requests flood the mailbox or until a trade contractor has worked on several houses before you discover that his or her work does not meet your standards.

Ensure 100-percent completion

You will no longer find that the company has paid for job that was supposed to be 100-percent complete, but was actually only about 90-percent complete.

Increase quality—decrease warranty

By following the scopes of work and using the inspection reports you hold a trade contractor to a high quality-of-work standard, thus eliminating the majority of warranty problems. By making the trade contractor complete a job 100-percent, you eliminate a large part of the punch-out on every jobsite, which also reduces personnel costs.

Save on wasted material

A few years back I received a phone call from a construction manager who was trying to convince his boss to set up the scopes of work program. The manager told me that he had tried everything he could think of and nothing was working. He asked me if I had any suggestions. I told him "Sure. Here's what you do. Gather up a hand full of material invoices, a pen, a pad, a calculator, and your boss. Go to one of your subdivisions and walk the houses and lots. Make a list of all the leftover and wasted material you find lying around. Then use the material invoices to assign a cost to each item. Add the items and see how much money is being thrown in the trash." I asked him to call me back and let me know the outcome of the trip. When he called me back later he told me his boss was sold on the program. The dollar amount of the wasted material was a real eye opener.

Several of the inspection reports require that a list of leftover material be created and turned into the estimating department. By utilizing these lists, the estimating department then can tighten up estimates and eliminate a large amount of wasted material.

Improve jobsite appearance

How many times have we all walked through houses and stepped over (or perhaps in) wet drywall mud, seen a bathtub full of liquid and trash, or hugged a wall because temporary handrails were not in place? I know we all have seen prospective buyers running obstacle courses of leftover building materials and debris both in and outside of a house. The scopes of work require all trade contractors to clean up after themselves. They are required to move all leftover material to a designated area and remove all trash to a designated trash site before the job is considered complete. Safety measures such as barriers and temporary handrails also must be in place at all times.

Very visible items, such as covered and protected tubs, cabinets, and countertops, along with clean yards and houses almost shout to potential buyers that this builder does quality work. When a buyer can see evidence of the care a builder takes with construction they believe the builder to be a quality builder. Whether a buyer realizes it or not, they equate a clean, neat jobsite and house to a good job, just as a dirty area will convey the message that the builder does sloppy work. Prospective buyers who feel a builder does quality work buy from these builders!

Provide legal recourse

When trade contractors sign the initial paperwork, they agree to perform their warranty or to allow the company to backcharge them for the cost of hiring someone else to finish the job. The terms and conditions require trade contractors to furnish proof of insurance and gives the company the right to file on the trade contractor's insurance, should it be necessary to fix a very expensive mistake.

3

Sounds Too Good To Be True.
What's the Catch?

By now you're thinking, "This sounds too good to be true. There has to be a catch." The catch? It's people, plain and simple. Many people resist change. They protect what they perceive as being their turf. They fear losing control of their job, or they are lazy and resist any extra work. Additionally, many employees also fear and resist accountability—accountability of any kind.

People in the construction industry are no different than people in other industries. Because of the nature of construction, however, their attitudes are sometime unusual. In most businesses, the people are all company employees and they expect to have to adjust to changes in the company's policies. This is not true of the construction industry. Employees in construction generally make up a very small percentage of the people involved in the building of a house. The majority of people involved in building a house are independent contractors or employees of a trade contractor.

As owners of their individual businesses, trade contractors feel they have the right to determine how a job should be done. Many trade contractors have been doing their job the same way for years and expect to continue to do it the same way forever. Like anyone, trade contractors can become resentful if they think a change is demeaning to their image or if they believe they are being asked to do additional work they consider unnecessary. Finally, there is a small percentage of trade contractors who routinely do a 90-percent job and are used to being paid for 100-percent completion.

Site personnel are only slightly different from trade contractors. Whereas most employees go to work for a company with the expectation of job security, the same is not true of jobsite employees. It is not at all unusual for site personnel to be laid off when a subdivision is complete, usually with no hard feelings on either side. Over time, many site superintendents develop the same attitudes found in trade contractors. They may view themselves as independent contractors, even if they are company employees. Typically, site superintendents have been doing their job the same way for years and don't really expect to have to make major changes in the way they run a jobsite just because they are working for a different builder.

The other thing that site personnel and trade contractors have in common is their aversion to and dislike of paperwork. The most prevalent attitude on any jobsite is that the field's job is to build houses, not do paperwork. Most field people will spend a lot of time and energy figuring out ways to avoid paperwork.

Many owners or top managers came from the field and many still feel more affinity to the field than to the office and paperwork. Yet management must be convinced of the value of the program and be prepared to support it 100 percent, or it will fail. It is difficult enough to sway the field staff and trade contractors without them knowing that top management won't enforce the program.

As I said before, the biggest hurdle you have to jump when implementing the scopes of work program is gaining the support of the people involved. You will have to convince everyone that the changes will benefit them as much as or more so than the company. It is a difficult process, but one that many companies have accomplished successfully. You can do it also. The benefits of the program, for everyone, far outweigh the time and effort it takes to make supporters of the program out of your people.

4

Top Management's In—
What Now?

Once you have the support of top management it's time to get the first part of the program underway. Administrative chores are the first item on the agenda and they usually take the longest amount of time to complete. The bigger the company the more time it will take because of the number of people that will be involved.

Preparing scopes of work from scratch takes several months and this is time you won't have to spend. I have done most of the work for you and it is included in the CD found in the back of the book. However, you will need to spend the time necessary to modify the samples from the disk to ensure that all documents in the scopes of work program fit your own business practices and all state and local codes.

Do first things first

Now it's time to get all the paperwork done and appoint a boss. You have six jobs to do before you can put the program into operation.

Job #1: Appoint a "boss"

The program needs a "boss." This person must be someone with enough authority to ensure that each department involved in the implementation of the scopes of work program performs the job required of them. This person must be able to devote the time to be sure that all steps are completed on time and correctly. Finally, the manager must have the authority to reprimand or terminate an employee or remove a trade contractor when that becomes necessary. Using one of the "clerks" in the office won't cut it.

Job #2: Print the documents

Print all of the scopes of work, terms and conditions, and all inspection reports from the CD included with this book.

Job #3: Modify the terms and conditions

The set of terms and conditions found on the CD includes two different documents. The document called Terms and Conditions covers what paperwork the trade contractor must furnish to the company. The document called General Terms and Conditions spells out what is required of anyone who works on the company's job or jobsite. The general terms and conditions is the one that should be printed on the back of all purchase orders so that they become a part of the purchase order. Both of these terms and conditions have to be reviewed to determine if changes are necessary to ensure they reflect what your company needs and wants. Once completed send both sets of the terms and conditions to your attorney for review to ensure they are enforceable in your own state or locality.

Job #4: Review and modify the Scopes of Work

Review each of the detailed job requirements in the scopes of work and make any changes, modifications, or additions needed to meet the demands of your particular business or the codes for your particular state or county. Determine if you need to add additional scopes of work for your region or for the particular way construction phases are handled in your area. Then review the pre- and post-inspection reports and make sure they match your final scopes of work.

Don't forget to change the inspection reports to match your scopes of work.

Job #5: Develop a scope of work for field positions

I always suggest that during this review process a company develop a scope of work for all field positions by defining the responsibilities of the field personnel in relation to the general scopes of work program. These documents should be developed in-house since each company's structure and expectations are different. You cannot expect your own staff members to perform their job if they don't know what is expected of them.

Job #6: Assemble the manuals

Once all of the necessary changes are finalized, full scopes of work manuals need to be made. A full manual includes a copy of both sets of terms and conditions, a copy of the scope of work for each construction phase plus the companion inspection reports, the scope of work for field personnel, and a copy of the company's policy for noncompliance. In other words, a full manual covers every step of construction from clearing and grading to the final cleaning, along with information for your employees, your trade contractors, and all the documents you ask a trade contractor or supplier to sign and the information for your own employees.

Manuals normally require a sturdy 2-inch, three-ring binder with tabs to separate each individual scope of work and its companion inspection report(s). The manual is easier to use when the scopes of work are alphabetized rather than placed in order of the construction process. The scopes of work for field positions and the company's noncompliance policy comes last. The number of manuals required will depend on the size of your organization and the number of different departments in your company. Prepare one full manual for each job shack, each project manager, the estimating department, the purchasing department, the warranty department, the department or person who signs up new trade contractors, and one manual for general office use.

Who's involved and what do they do?

To accomplish the six jobs I listed above you will have to assign certain jobs to certain people or departments. The following will give you an idea of who will be involved and what their jobs will be. If you are a builder who does not have all of the departments I talk about don't think the job is too large and give up now. Review what needs to be done and adjust the jobs to the people and time you have.

The clerical staff

Select the person(s) who will be responsible for the clerical side of the project. The implementation will go more smoothly if the same people are involved from start to finish and continue to be responsible for the paperwork in the future. In the beginning, their jobs will be printing the documents from the CD found in the back of this book, making the necessary number of copies of the documents, collecting the documents after changes are made, typing the modifications, printing the final copies, preparing the sign-up packets for the trade contractors, and preparing the final scopes of work manuals.

During day-to-day operations, the clerical staff will be responsible for ensuring an adequate supply of inspection reports for all jobsites and sign-up packets. As you begin to use the scopes of work program you most likely will find it necessary to make small changes to the scopes of work and the inspection reports. The clerical staff will be responsible for making the changes and then distributing updated copies for all manuals and mailing updated copies to trade contractors.

You will need to decide how the inspection reports will be distributed to the trade contractors and to the field. It is not critical that trade contractors receive a copy of the inspection report for each job. A copy of the inspection report that will be used for their phase of construction will be included in their sign-up packet, and unless the inspection report changes, they don't necessarily need a new copy for each individual job. If your company produces preprinted purchase orders it is relatively easy to attach an inspection report to the purchase order. Whether or not you give the trade contractor a copy of the inspection report is a decision you must make based on how you do business and the size of your staff.

For larger trade contractors, who send different employees to the jobsites, it keeps confusion to a minimum if an inspection report is given to their employees before they leave for the jobsite. I have found that the larger trade contractors like to have a supply of inspection reports in their office so they can include one with the job orders they issue their employees.

The field, on the other hand, must have an inspection report for every phase of construction for every house under construction. The easiest way to handle the field's inspection report is to keep an ample supply of inspection reports in the job shack. Then the site superintendent will always have a copy of a report when one is needed. In my day-to-day working with the scopes of work program I have found the standard excuse for not completing an inspection report to be, "I didn't have one." Trade contractors also can pick up copies from the job shack whenever they need or want a copy. The success of this method often depends on whether or not there is room in the job shack for a small file cabinet. Another method is to send a full set of inspection reports to the field along with the plans and start requests. Your company may have a different idea that will work better. It really doesn't matter how the inspection reports get to the field as long as they are there.

The clerical staff also will be responsible for preparing the initial sign-up packets for trade contractors and ensuring that an adequate supply of them are on hand. The initial sign-up packets should include two copies of: both sets of terms and conditions; the scope of work pertaining to the contractor's particular trade; the companion inspection report(s); and, a stamped envelope addressed to the company. Place the documents in a large envelope and write the trade contractor's name and trade on the outside.

These same items, excluding the stamped, self-addressed envelope, form the basis of the sign-up packet. You can add any other forms to the packet that your company uses when they sign-up a trade contractor.

Technical experts

Those in (or outside) of the company with technical construction knowledge are the people who should review the scopes of work and determine any changes or modifications that need to be made. I have tried to be sure the scopes of work found in this book and on the enclosed CD all conform to the NAHB Performance and Building Standards and Residential Construction Performance Guidelines. However, every state and county has its own individual code requirements, and you should incorporate these into your scopes of work. The scopes of work also should be reviewed against your company's own building standards and modified to include your requirements. There are always regional differences in the phases of construction. Additional scopes of work should be developed for those differences, or existing scopes should be combined or split.

The purchasing and estimating departments

The scopes of work may make it necessary for the purchasing department or estimating department to change their paperwork or procedures. Give each department a working copy of the scopes of work and inspection reports for review. There may be procedures in both departments that will need to be included in the final scopes of work or terms and conditions.

Material or supply items printed on your purchase orders may need to be modified or expanded, or wording may need to be changed to match the detailed job requirements of a particular construction phase. Print the general terms and conditions on the back of every purchase order to make them a part of the document. There is a set of terms and conditions for trade contractors. The enclosed CD includes a sample purchase order for trade contractors and another for suppliers.

The estimating department should work out the procedures for collecting the extra material lists from the field, and the procedures for modifying existing estimates based on the information coming in on these lists.

The trade contractor coordinator

The trade contractor sign-up process will need to be modified to include having the trade contractor read and agree to both sets of terms and conditions and the scope of work for their particular construction phase and the companion inspection reports. The sign-up documents require the signature of the trade contractor and a company representative. Two sets of documents must be signed. Give one set of documents to the trade contractor. The other signed set is kept in-house.

A signed copy of a trade contractor's scope of work papers should be filed and kept in case it ever becomes necessary to remind the trade contractor of the terms and conditions they agreed to abide by. In extreme circumstances, your attorney will need these documents. For example, suppose a trade contractor who no longer works for you doesn't want to do his or her warranty. The signed document states that the trade contractor agrees to perform his or her warranty or the company may charge the contractor for the cost of having someone else perform the work. These signed documents may help you if legal action becomes necessary.

The person or department responsible for signing up your trade contractors must become familiar with each and every scope of work to be able to answer questions and explain the program in detail. They must be very familiar with each scope of work. These employees must have a complete, up-to-date copy of the scope of work manual that they can refer to and use to make copies for sign-up packages.

The accounts payable department

Each company has its own system of collecting bills from trade contractors and suppliers. Some have a meeting each week during which the field people turn in invoices, while other companies require that all invoices be mailed to the office. Some companies use a combination of the two. Regardless of how your company collects bills, the accounts payable department should be trained to review each invoice, purchase order, and the corresponding inspection reports to ensure that everything is filled out correctly and signed off on. If an inspection report is missing or not signed and dated, the accounts payable staff should deny payment. The trade contractor should be notified that they are being denied payment, and the bill and inspection report should be given to the person responsible for getting the problems corrected.

Upper management and human resources

You must be prepared for those people who ignore the program. Your company will need to develop its own policy on what it will do to those who don't follow the program. It is always best to involve top level managers and someone from the human resource department in the development of this policy. The best policy in the world is worthless if the enforcement does not begin at the highest level. So, save yourself some grief later and make sure everyone is in agreement on what actions the company will take for noncompliance.

Here is the noncompliance policy my company used. You may want to use it as a guideline when you develop your own. The first time an employee fails to follow the procedures detailed in the scopes of work, inspection report and/or the general terms and conditions, the person receives a verbal warning with the person's direct supervisor present. A note that the warning was issued verbally should be placed in the person's personnel file. For the second offense, the employee receives a written warning and a two-day suspension without pay, again with the person's direct supervisor present. A copy of the warning is placed in the individual's personnel file. The third offense results in termination. This may seem harsh, but if you do not make believers out of your field personnel the program will never work.

The same steps are applied to trade contractors. For the first offense notify the trade contractor in writing that he or she stands to be placed on the Do Not Use List. On the second offense, take away the trade contractors' next job in line. For the third offense, put them on the Do Not Use List and remove them from your jobsites.

5

Take a Deep Breath and Let's Get Underway!

You are now ready to begin implementing the program. Run a final check to be sure everything is ready before you move to this phase. Check that:

- Terms and conditions have been modified and approved by your company attorney.

- The detailed job descriptions for each construction phase have been finalized.

- The scopes of work are finalized and include all items found in the detailed job description.

- Each scope of work has been checked against your own state and local codes.

- There is a scope of work for every phase of construction.

- There is an inspection report(s) for each scope of work.

- The inspection reports have been checked against the detailed job descriptions.

- A scope of work has been completed for field personnel.

- A written policy stating the consequences of noncompliance with the program has been finalized.

- Manuals have been prepared and are ready for distribution. Each manual should contain two copies of the general and specific terms and conditions, a copy of each scope of work, a copy of each inspection report, a copy of the field staff's' scopes of work, and a copy of the company's policy for noncompliance.

The paperwork is all done and now it's time to deal with people. Unless your company is very small it is more likely that during the preparation of the paperwork rumors about the program were flying fast and furious. If the word about the program has gotten out everyone is curious about the program and many people are prepared to hate it - sight unseen. Let's take a look at what you can expect during the implementation of the program and some of the ways to present the program to your people.

Anticipate people's resistance

In chapter 3 I said that the biggest problem in implementing the scopes of work program is people. Be prepared for resistance. Depending on the relationship between the people holding the implementation meetings (described below) and the attendees, you may or may not get vocal resistance, but you will most certainly get silent resistance.

I have never implemented this program without a lot of resistance from the field personnel and trade contractors. If you are implementing this program during a tight construction market, you can be heavy handed and demand that employees and trade contractors follow the program. However, when the market becomes hot again, you may lose most of your people. It is much better to try to overcome the initial resistance and convince people that the program will benefit them as well as the builder (which it does).

Now take a deep breath and let's get the implementation under way. I'll give you some helpful hints as we go along.

Get things underway

Rather than talking to everyone individually, it's best to hold meetings for the different groups affected by the program. Be sure that the person holding the meeting knows the program inside and out.

Convince your field staff

Your first step is to set up a meeting with the field staff, including site superintendents, project managers, and punch-out people. Begin your meeting by explaining why the company is implementing the scopes of work. (All of these reasons are listed in chapter 2.) Don't expect the field people to be terribly impressed with these reasons, but the idea is to let them know why the company chose to use this program.

Then explain how the program will make their job easier. Here are some of the things I use when I hold this meeting:

- The program gives them a checklist to use for every step of construction. The checklists will help them spot mistakes or sloppy work that might otherwise go unnoticed. Remind them that everyone is human and we all can overlook things.

- The use of the checklists will eliminate problems for the next trade contractor in line who must spend extra time or materials trying to work around mistakes or sloppy work.

- They will no longer need to spend time trying to get a trade contractor back to fix a problem or, worse yet, have to take them off a current job to go back and fix a mistake at the previous job.

- Construction schedules will not be held up waiting for trade contractors to come back to fix something or to complete the job.

- The program will ensure that the job is ready for trade contractors when they show up for work. Every site superintendent knows that if trade contractors show up ready to work and the job is not ready for them, they will go work somewhere else. They won't come back until they finish the other job, which can wreak havoc with a construction schedule.

- The program will attract a better class of trade contractors to the jobsite, making the site superintendent's job easier.

- It will help eliminate the last minute scramble to bring a house in on time (which always results in a huge walk-through list).

- For the most part, it will eliminate irate homeowners stopping by the job shack to complain about a warranty item.

Once you have explained how the program will help them, give everyone their copy of the scopes of work manual and tell the site superintendents that they should find a spot for the manual in the job shack.

At this point, many field personnel take offense and assume you are suggesting they need a book to know how to build houses. Restate that the manual is intended to be a reference tool. The manual should be used whenever there is any doubt about whether a job has been performed to the company's standards or to settle a question about how a job should be done. Referring to the scopes of work often puts a quick end to an argument between a site superintendent and a trade contractor. I usually point out to the field people that by using the manual during a dispute, office and management personnel can be blamed and they can sympathize with the trade contractor to help keep a good relationship intact.

Go over the terms and conditions. Be sure everyone understands them and understands that the field personnel are expected to enforce them.

Ask everyone to look at the first scope of work. Go through the document carefully. Look over the detailed job requirements, then the scope of work, and then the inspection report(s). When you come to the inspection reports you will either get (if your people are vocal), "I don't have time to do this and build houses too," or (silent resistance) you will see a lot of people slide back in their chairs and cross their arms. Inspection reports are the major stumbling block for field people.

At this point, I usually ask the question, "Is every house you build one you would be proud to live in?" To date, I've never had anyone answer yes. I usually follow up with, "Why not? Are you telling me you would build your own home with more care than you build the ones we sell? Are you telling me you really don't check the work in each house at each stage of construction?"

Remind the field personnel that their job is much more than scheduling and ordering materials. Their primary job is to watch every step of construction, on every house, to ensure that the house is being built per plan and to the company's standards. If the field personnel are doing their job, they are already checking each stage of construction. Using the inspection reports will not add that much time to the inspection process. Remind the field people that they are human and can overlook something. By using the inspection reports, they won't forget to check everything.

End the meeting now. Suggest that each person take the manuals home with them, look them over, and be prepared to discuss them in detail at the next meeting.

Begin the next meeting by asking everyone what they think of the scopes of work. Ask if anyone has found anything that needs to be changed. Then ask enough pointed questions to determine if everyone has read the manuals. If not, you'll need to review each scope of work individually to be sure everyone is aware of what is required. If you don't do this, you will eventually hear, "Well, no one told me."

In particular, review the scopes of work for the field personnel. Don't give anyone the chance to wear out the "Well, no one told me" excuse. This meeting is your chance to explain the consequences of noncompliance with the program. Explain what will happen to both employees and trade contractors who don't follow procedures. Let everyone know what is expected and what will happen if they do not follow the program.

Get buy-in from warranty and quality control

The warranty department and, if the company has one, the quality control department are next on your meeting list. All of the requirements in the scope of work program will impact these departments. Begin this meeting the same way you began the meeting with the field staff. Explain why the company chose the program and then what the program will do for the warranty staff. The biggest impact on the warranty staff will be the reduction of warranty , particularly expensive, time consuming warranty. I have found that, out of everyone connected with a company, the warranty department welcomes the scopes of work program and often become its greatest supporters. It doesn't take very long for the warranty staff to understand that this program will eliminate their biggest headaches and reduce the number of homeowners who are impossible to satisfy.

Most companies with a separate warranty department also use these staff members for walk-throughs and orientation. These employees, more than anyone, know that the homebuyers' unhappiness grows in direct proportion to the size of the walk-through list. A huge walk-through list gets the warranty department's relations with the homeowners off to a bad start, and they must deal with these homeowners over the next year.

Most warranty departments are responsible for trying to get trade contractors to perform the warranty for which they are responsible. In larger companies one full-time person often is devoted to this chore, which can be a vicious cycle. The longer the trade contractor puts off doing his or her warranty the longer homeowners have to wait for the problem to be corrected. The longer the homeowners wait, the more unhappy they become. At the same time these homeowners become more vocal to the warranty department. Eventually, the homeowners begin to tell their friends, family, and coworkers about their unhappiness with the builder. Under the scopes of work program, the trade contractor who refuses to do his or her own warranty work can be backcharged for the cost of someone else repairing the problem. With the scopes of work program warranty issues can be addressed quickly, breaking this negative cycle.

At the meeting with warranty and quality control personnel, briefly go through the manual and talk about the scopes of work and the inspection reports. The most important thing for the warranty and quality control departments is that each person understands who is responsible for what, how the trade contractor is to be contacted to handle warranty, and where a copy of the trade contractor's contract can be found. Explain who should be notified and how to report items found on the job that are not in compliance with the inspection reports and who should be notified.

Inform your trade contractors

The third group you need to meet with is your trade contractors. I have found the best way to meet with all of your trade contractors is to hold one large meeting. I usually hold my meeting the day before trade contractors normally are paid, and I tell them they may pick up their checks at the meeting (as an added incentive to attend). Give serious thought to where and when you will hold the meeting. Set the meeting for a time and place as convenient as possible for the trade contractors. I know a builder who has an office located on the very outskirts of a large metropolitan area. This builder always holds meetings at a hotel close to the office at 2:30 in the afternoon. Many trade contractors must travel more than an hour to reach the meeting and then must face rush hour traffic to get home. The trade contractors arrive at the meeting unhappy with the builder. Don't make the same mistake.

Be prepared to talk about the ways the program will benefit the trade contractors. The first thing I always stress to trade contractors is the fact that they are in business, just like the builder, to make money. Many of the things that waste time and money for the builder are the same things that waste time and money for trade contractors. Trade contractors don't want to be called back to a job for a small repair, or arrive at a jobsite ready to work only to find the job isn't ready. Unless a builder is very large, trade contractors usually work for more than one builder and must schedule their work to meet the demands of two or more builders.

Here are a few points you may wish to make to the trade contractors.

- The detailed job descriptions and scopes of work protect the trade contractor. If they perform their job to these standards, they will not be called back to repair or finish a job.

- If the trade contractor is large enough to have employees, they do not have to worry about their employees not completing a job correctly.

- The time and money the trade contractor spends on warranty work will drop dramatically.

- Trade contractors are guaranteed that their work area will be clean and ready for them.

- Through the use of the inspection reports the trade contractor is protected from being accused of or charged for damage they did not do.

- The inspection reports ensure that trade contractors will not have to work around the errors made by prior trade contractors.

- Trade contractors are guaranteed payment, on time, as soon as the job is completed. There will be no more backcharges by field personnel.

- The number of times a trade contractor can be called back after a walk-through is spelled out. Once that number is passed, the trade contractor can bill the company for their work.

There will be other things that come to mind that you might want to use. Know your trade contractors' major gripes, and be prepared to address those issues.

You also need to explain that there is a trade-off here. The trade contractor must complete the job to the company's standards before they get paid. The job must be 100-percent complete. They won't be able to ask the site superintendent to turn in a bill on a job that is not complete, promising to finish it later. There will be things in the detailed job descriptions and scopes of work that the trade contractor will not like. I have found cleanup to be the number one gripe. It seems that trade contractors don't want to clean up after themselves, but they are the first to complain if the job area is a mess.

Now, open the meeting to questions. Do your best to answer them all. If you are not able to answer something, promise to find the answer and get back with the person. Then be sure you do. You are trying to build credibility. Don't start off on the wrong foot.

Next talk about the sign-up packet. Use a sample to explain each item. Be clear about what is in the packet, how the paperwork is to be returned to the company, and what the consequences are if the paperwork is not returned on time. For example, explain that the trade contractor should sign both sets of documents and return both sets to the company using the stamped, self-addressed envelope. Tell them that a company representative will sign both sets and return one signed set to them.

Do not ask the trade contractors to sign their paperwork at the meeting. They must be allowed time to read and understand what they are signing. I always place the sign-up packets on a table at the back of the room (with their paychecks), and ask the trade contractors to pick up their packet and checks as they are leaving. Be prepared for questions from trade contractors after they have had time to read the information.

As a suggestion, I have found that a follow-up meeting is very effective. Schedule the meeting after the scopes of work and inspection reports have been in use for a month or two. At this meeting allow the trade contractors to express their problems in the scopes of work, inspection reports, site personnel, or other issue they found with the implementation. I also ask for suggestions for improving the program. Some of my best suggestions have come from trade contractors. By addressing the problems that come up during in this meeting you can be sure the scopes of work program will enjoy a long life.

Notify suppliers

You also might want to have a meeting with all of your suppliers, or you may mail a copy of the terms and conditions with a cover letter. A meeting with suppliers usually will be short because they are concerned only with the company's terms and conditions as they relate to them. Pass out a copy of the terms and conditions and go over each item on the form. If your company uses preprinted purchase orders, explain that the terms and conditions will be printed on the back of the purchase order and will become a part of the purchase order. The majority of the people at the supplier's meeting will be salespeople who do not have the authority to sign anything on behalf of their company. They will need to take a copy of the form back to their office for review and approval.

6

Call in the Enforcers!

As with all new things, day-to-day operations are the most difficult part of the program. I have already said that field personnel and trade contractors won't expect to have to abide by the program. This is the time to review the scope of work you prepared for your field personnel and your policy on noncompliance with the program.

It would be wonderful if, after all of your hard work, the program worked from the start and everyone in the field followed the procedures and completed the inspection reports like they are supposed to. I've never seen it happen. With the majority of site superintendents, my experience has been that trade contractors drop by the job shack to turn in their bill and the site superintendent pulls the inspection report for the job from the file. Without leaving the trailer, the inspection report items are quickly ticked off on the report. The field people assume no one will ever know how or when the inspection reports were completed.

The Boy Scout's motto is "Be Prepared." Adopt it for your own. Select a group of people (the number of subdivisions in operation and the number of houses under construction will determine the number of people needed) who are knowledgeable about construction. For the next few weeks, this group's responsibility will be to compare the inspection reports that come in form the field to the actual job performed by the trade contractor to determine if the job was completed per the company's standards and the inspection report.

At this point in time, don't choose project managers (or whatever title you give the person responsible for the operation of two or more subdivisions and the supervisor of site superintendents). If you have a warranty department, use the warranty manager(s) and staff. If you have a quality control department, use them. If any top managers have field experience, sign them up also. Choose people who have more loyalty to the company than to the field.

If your company is like most, there is no such thing as a secret. After you choose your group, don't explain what you want out of them. Just have them set aside time on a particular day to "do a job." If you tell anyone what you have in mind, trust me, the field personnel will know within an hour and your purpose will be defeated.

The following examples use the consequences for noncompliance outlined in chapter 4. I am aware that not every company has the number of departments, or the staff, used in these examples. Modify the procedures to fit your company and your staff.

Get through week one

When the first group of invoices and inspection reports come in, gather them up and distribute them to your group instead of forwarding them to the accounts payable department. Remember the idea is to take the inspection reports, go to the subdivision, and check every one of them against the actual job. My experience has been that you probably will find that only a very few of the reports are correct.

Next, notify the project managers and/or site superintendent of the problems your group found. Return the bills and the inspection reports and issue a verbal warning to the site superintendent. Follow through by notifying the trade contractors of the problems your group found. Tell them that payment of their invoices is being withheld pending correction of the problems. Also, give them their first warning.

Play hardball in week two

During the second week follow the same procedures. This time, however, give the inspection reports to the project managers and have them check the jobs. When the project managers return the inspection reports, give them to your group and send them back out to recheck them. Don't be surprised if you find that your project manager didn't report omissions or incomplete work. What the project manager probably did was to tell the site superintendent, "Hey get this fixed right away. I'll cover for you." Issue a verbal warning to the project manager and a written warning to the site superintendent if this is the second time the site superintendent's inspection reports were found to be incorrect.

Once again you must follow through by notifying the trade contractors of the problems and telling them that their invoices won't be paid until the problems are corrected. If this is the second offense for a trade contractor, remind him or her of the consequences of noncompliance with the program.

Maintain consistency in week three

I wish I could tell you that by the third week everything will come in correct, but I can't. Usually by the third week about half of the site superintendents have gotten the message and are checking jobs and completing the inspection reports with the trade contractors. The other half, however, are breathing a sigh of relief and thinking, "Whew that's over!" So one more time, gather up the inspection reports from the field and give them to the project managers to check. When the project managers return the paperwork, once again gather up your group and recheck the reports. You most likely will have to issue a written warning to at least one project manager and fire one site superintendent.

Now you have the attention of everyone in the company. The message is, "The company is going to enforce this program."

For the next few weeks, invoices and inspection reports generally will be correct and then the field will start getting slack. Be prepared to "spot" check inspection reports indefinitely. It is the only way to ensure compliance by everyone, all the time.

You probably will hear from trade contractors about field personnel or project managers who are not following the rules. The good trade contractors aren't interested in losing their job with a good builder. Pay attention to the gripes from the trade contractors because they probably went to the project manager first. The fact that the project manager is not handling the situation should be a warning flag that you may have an employee problem.

This Had Better Be Worth It!

Do I hear you saying, "Boy, this better be worth it?" To answer your question, as a matter of fact it is. Here are the reasons why.

Where did the warranty go?

You won't see an immediate drop in warranty, but if you are interested in seeing just how well the program works, track the warranty on the houses that were started **after** the third week of the program's operation. Compare the warranty on those houses to the warranty on prior houses. You will be astonished at the drop in warranty requests and the cost of repairs.

Small, cosmetic warranty items do not cost the company a lot of money. Rather, it is major warranty items like the out-of-square foundation, the brick lintel installed too short and bricks that are sliding off the house, sagging cabinets, out-of-plumb doors and windows—there is no need for me to go on. You are fully aware of the warranty nightmares and what they cost. By eliminating these high-cost items, you put money in your pocket.

Talk to the warranty department and/or the person who takes the phone calls from irate homeowners. You should find they are receiving fewer and fewer phone calls. Less time spent listening to irate homeowners is time spent more productively on other things.

Increasing sales

Talk to your sales force and see what kind of comments they are getting from prospective buyers. You should see an increase in sales. It's difficult to attribute increased sales to any one thing, but clean, neat subdivisions and visible signs of care (covered bath tubs and countertops, for example) often influence buyers' decisions to choose your company over another.

Showcasing quality

Walk through your subdivisions and look at the quality of the work being done on your homes. You will see the difference. Watch and listen to prospective buyers. You will notice that they stay longer looking at your homes. You also will hear comments about how nice the community looks.

Saving on excess materials

If the estimating department is doing its job with the excess material lists you should see a small decrease in the hard cost of all houses. In a volume business even a savings of $50 to $100 on every house adds up quickly.

Reducing walk-through lists

Look over the walk-through lists for completed houses that were started after the third week that the program went into operation. Compare those lists to the walk-through lists of houses completed earlier. The lists should be significantly shorter. The items should be mostly cosmetic, with few serious problems.

Making best use of employees

If you have had a quality control department in the past, it should be much smaller or no longer be needed. At a minimum, you should need fewer employees in the department and their responsibilities should primarily be "spot checking" inspection reports. We all know that it's impossible to eliminate the warranty department completely. The number of employees needed in this area also should be greatly reduced. Depending on the size of your subdivision(s), the number of punch-out and clean-up positions also should be greatly reduced, if not eliminated entirely.

Evaluating employees

Instead of finding out that a field superintendent is a marginal or inept employee after one or several houses are complete, the program enables you to weed them out before too much damage can be done. This also is true for trade contractors. It's almost impossible to attach a dollar figure to the cost of a bad site superintendent or trade contractor, but the damage they do reaches the bottom line one way or another.

Improving the bottom line

And last but not least, check out your bottom line. That's where the final proof will be found. There is no way I can tell you the amount your bottom line will increase. The results will depend on your company's size, the size of your staff when the program started, and the number of houses you build in a year. I can tell you that with every implementation I have been involved with, the bottom line has increased significantly once the program was put into day-to-day operation.

Appendix A

Sample Scopes of Work

The following scopes of work can be found on the enclosed CD. I suggest you print these scopes off of the CD for review by your technical and management staff, and by your company's attorney where appropriate. See pages 231–232 for the scopes of work and its corresponding file name on the CD, as well as information for using the CD.

I have tried to make the scopes of work as nonregional as possible. There are some scopes that include region-specific specifications for clarity. For example, in the scope of work for insulation I have left in the R-values of my area so that you may see how the scope of work and the inspection report should be formatted and what it should include.

While it is impossible to include a scope of work for every region of the country, I have tried to include the standard construction phases that are found almost everywhere. If there is not a scope of work for a particular job or trade that you need, follow the standard format of the scopes of work and inspection reports to produce your own. You may need to split some scopes or combine others to fit your particular area or business.

Review each of the scopes of work by printing them from the CD or copying the printed lists in the book. Make the changes required for your region, your local codes, or the specific requirements of your company. The scopes are formatted on the CD so that you can add a logo or company name at the top. Using the search and replace feature will allow you to change the words "The Company" to your own company name or initials. The first page of all the scopes of work is standard. If you wish to make modifications to this first page, change one page and then copy-paste your modified page into each scope of work document. All "Detailed Job Requirements" sections found in the scopes of work are numbered using Microsoft Word's®automatic numbering system. This will allow you to add or delete items without having to re-number the list by hand. This function also will allow you to split or combine scopes to fit your area or company procedures.

<u>SCOPE OF WORK</u>
Standards and Description of Work Performance

BLOCK FOUNDATIONS

The Company's **Terms and Conditions** are by reference a part of all **Scope of Work** requirements.

Construction Requirements:

Generally speaking the work of Trade Contractors and their employees is expected to be performed in a good and workmanlike manner. Workmanlike quality is defined as workmanship that meets or betters those criteria indicated in applicable building codes, using materials and installation methods identified in the construction plans and this Scope of Work.

Code Requirements:

All jobs shall conform to those standards stipulated in the building code, mechanical code, plumbing code, and electrical code applicable in the local jurisdiction. All construction on The Company's jobsites shall meet or exceed NAHB Performance and Building Standards.

General Comments:

The Company considers our Trade Contractors to be experts at producing a high-quality job. Everyone on our construction team—staff, Trade Contractors, and suppliers—however, must recognize the importance of providing quality in both the product and service areas while on our jobsites and in the homes of our purchasers.

Since we work as a team, poor quality or service, from any of us, reflects unfavorably on all of us. An exceptional level of product quality and highly effective service can help us all to increase our business and grow.

The Company's definition of quality construction also requires that every job be completed correctly the first time. When this does not occur it costs both of us additional money, imposes on the purchaser, and hurts our reputations as quality builders. That is why, in situations where construction was not completed in a quality manner, prompt corrective action is required to remedy specific deficiencies.

In the following information the term Site Superintendent shall refer to any The Company representative with authority to perform the specified task. The term Trade Contractor shall mean the Trade Contractor's organization or any representative that is assigned the authority to perform the specified task.

General: Masonry work can be performed with a variety of materials, methods of application, and installation techniques. This range of choice creates a distinctive situation that can vary with each installation. More than any other trade, masonry relies on the skill of the individual for the quality of the final product.

Company Rep's Initials _____
Trade Contractor's Initials _____

Masonry foundation walls should maintain structural integrity and not crack in excess of 1/8 inch in width or displacement. Foundation variance should not exceed 1/2 inch out of level in 20 feet, with no ridges or depressions in excess of 1/4 inch within any 32-inch measurement. Foundation walls should not be more than 1 inch out of level over the entire surface and not vary more than 1/2 inch out of square when measured along the diagonal of a 6x8x10-foot triangle at any corner.

Masonry should be laid within 3/8 inch of plumb, level, and true to lines in 4 feet. The overall plumb, level, and wall trueness variance shall not exceed 3/4 inch in 20 feet. Masonry or veneer walls should not crack in excess of 1/4 inch in width or displacement.

Materials: Specified materials should be handled and stored properly prior to construction. Masonry units should be covered and stored off the ground. Mortar should be suitable for the intended use and exposure. Cementitious materials should be kept dry. Concrete masonry units should never be wetted before and/or during laying in the wall.

Installation: All work is to be done by trained, experienced individuals. Mortar should be proportioned, mixed, and applied to ensure full mortar joints. Mortar should be protected from water penetration and efflorescence and should not be allowed to freeze during construction. Mortar is workable when it spreads easily, holds the weight of the masonry unit, and clings without dropping off or smearing.

Masonry units should be moved to their final position while mortar is still pliable. Exposed mortar joints should be thumbprint hard before striking with a round or v-shaped joiner. Mortar should not be smeared into masonry-finished surfaces.

Upon completion of the work all masonry should be carefully cleaned. When acid cleaning is required mortar should be well hardened and not less than seven (7) days old. The acid solution should be tested in an inconspicuous place to ensure against adverse affects.

Holes left by nails or line pins should be pointed while adjacent mortar is green. Where flashing is installed, masonry surface should be smooth and free from projections.

Payment: The quoted price per block is the standard price and does not depend on the size or type of block. Headers, 8-inch, 12-inch, caps, etc., all are paid at the same amount per block. Stucco is paid by the square foot.

Preprinted purchase orders are made on a "best-guess" estimate as to the number of blocks required. At the completion of the job, the Site Superintendent and the Trade Contractor together must field count the actual blocks used in the construction of the foundation. Both the Trade Contractor and the Site Superintendent must sign the field count. If the field count is not included with the Trade Contractor's invoice the invoice will not be paid. The office copy of the purchase order must reflect the number of unused blocks and the quantity of all other unused materials such as rebar, Portland cement, etc.

Warranty: The Company believes that all work done in connection with one of our homes should be of quality work and that all Trade Contractors should stand behind the quality of their work. Therefore we require all Trade Contractors who install foundations to warrant the quality of their

_____ Company Rep's Initials
_____ Trade Contractor's Initials

work for a period of one (1) year from date of closing of the house. Please refer to our printed Limited Warranty booklet for specific items that are covered under Warranty.

The Trade Contractor shall have seven (7) days in which to correct any Warranty problem. If the problem is not corrected within seven (7) days, The Company shall correct the problem and backcharge the Trade Contractor at the rate of $25.00 per hour, with a minimum charge of $100.00 plus the cost of any materials.

Inspection Reports: The Trade Contractor and the Site Superintendent shall walk the job together and complete each section of the inspection report(s). The Trade Contractor must correct any deficiency found during the inspection and the job must be 100-percent complete before payment will be made. The Trade Contractor and the Site Superintendent must sign off on all sections of the inspection report(s) attesting that the job is 100-percent complete and is correct per the job requirements found in this Scope of Work.

Detailed Job Requirements:

1. A new set of plans is required for each house. Plans are subject to changes and modifications. It is the responsibility of the Trade Contractor to have the new plans before beginning work. Plans should be picked up from the Site Superintendent. The Trade Contractor at no cost to The Company will correct any errors that occur from using an incorrect set of plans.

2. Purchase orders should be picked up from the Site Superintendent.

3. The Trade Contractor and the Site Superintendent must walk the job together and complete the pre-work section of the inspection report(s) before work may begin. Both parties must sign-off on the pre-work section of the inspection report(s).

4. Foundations shall be built per plan.

5. Masonry units shall be 8x8x16 CMUs. Header blocks shall be used for all slab foundation areas. Cap blocks shall be used for all areas of the garage and/or split-level walls.

6. All units are to be fully mortared with no spacing.

7. Masonry units should be moved to their final position while mortar is still pliable. Exposed mortar joints should be thumbprint hard before striking with a round or v-shaped joiner. Mortar should not be smeared into masonry-finished surfaces.

8. All foundation walls higher than 36 inches require that every third cell be concrete filled.

9. The block foundation Trade Contractor is responsible for supplying and setting anchor bolts. Bolts shall be 1/2-inch-diameter anchor bolts placed 6 feet on center and not more than 12 inches from corners. Bolts shall extend a minimum of 15 inches into masonry or 7 inches into concrete.

10. All walls shall be true, plumb, and square. Foundation variance should not exceed 1/2 inch out of level in 20 feet, with no ridges or depressions in excess of 1/4 inch within any 32-inch measurement. Foundation walls should not be more than 1 inch out of level over the entire surface and not vary more than 1/2 inch out of square when measured along the diagonal of a 6x8x10-foot triangle at any corner.

Company Rep's Initials _____
Trade Contractor's Initials _____

11. Masonry should be laid within 3/8 inch of plumb, level, and true to lines in 4 feet. The overall plumb, level, and wall trueness variance shall not exceed 3/4 inch in 20 feet. Masonry or veneer walls should not crack in excess of 1/4 inch in width or displacement.

12. All foundations must be protected from frost. During conditions when frost may occur, all block foundations must be covered with polyethylene. The polyethylene must be adequately anchored to prevent it from blowing off.

13. All excess mortar shall be removed from the sides of blocks both inside and outside. The foundation shall be smooth and ready for stucco.

14. Stucco shall be a uniform **minimum** 3/8 inch of Portland cement from footing to finish grade with no "thin" or "missed" spots, no runs, lumps, bumps, etc.

15. All excess blocks shall be stacked in one area at the front of the lot. All excess Portland cement shall be stacked and covered with polyethylene.

16. All trash and debris shall be removed from lot to dumpster or area designated for trash.

17. All runoff concrete/Portland cement shall be removed to the driveway area; it is not to be left on the ground.

18. A field count by the Site Superintendent and the Trade Contractor together is required prior to approval of the Trade Contractor's invoice for payment. Both must sign the Trade Contractor's invoice certifying that the count is correct.

19. The Trade Contractor and the Site Superintendent shall count the unused blocks and the quantity of all other unused materials such as rebar, Portland cement, etc. This information must be listed on the bottom of the purchase order.

20. The Trade Contractor and Site Superintendent must walk the job together and perform a final inspection of the job. The final section of the inspection report(s) must be completed and signed-off on by both parties. The inspection report must be attached to the office's copy of the purchase order and the Trade Contractor's invoice or payment will not be issued.

21. Any items found during the final inspection that need correction shall be corrected before payment will be made.

I _____ agent for _____

_____ have read and fully understand the above **Scope of Work** and I hereby agree to perform all work in accordance with the above.

Date: _____   _____
                             Signed: Trade Contractor (or agent)

Date: _____   _____
                             For The Company

INSPECTION REPORT

BLOCK FOUNDATIONS

Subdivision/Lot # _____

Pre-Work Inspection: (to be completed prior to beginning work)

____ Plans have been picked up and signed for.

____ Copy of purchase order has been picked up.

____ Work area is clean and free of debris.

____ Footing(s) are poured. Alignment and square corners have been verified.

____ Dimensions are correct.

____ Stakes are set and locations are correct.

____ Straightness (alignment) of lines is verified.

____ All setbacks, bay windows, fireplace(s), porches, stoops, etc., are clearly marked and dimensions checked.

____ All materials are on site and ready to be used.

Date: _____   Site Superintendent _____

Trade Contractor _____

Final Inspection: (to be completed before Trade Contractor leaves jobsite)

____ Verify all walls are true, plumb, and square.

____ There are no cracks in excess of 1/4 inch in width or displacement.

____ The variance does not exceed 1/2 inch out of level in 20 feet, with no ridges or depressions in excess of 1/4 inch within any 32-inch measurement.

____ Foundation walls are not more than 1 inch out of level over the entire surface and do not vary more than 1/2 inch out of square when measured along the diagonal of a 6x8x10-foot triangle at any corner.

____ Header block are used for all slab areas.

____ Cap block are used for all garage and drive-under walls.

____ Every third cell is filled in all areas of the foundation that are more than 36 inches in height.

____ All lines and pinholes are filled.

____ All excess mortar between blocks has been removed.

____ Stucco finish is smooth and a minimum of 3/8 inch of Portland cement parging is applied from footing(s) to finish grade.

____ Anchor bolts are 6 feet on center and not more than 12 inches from corners.

____ The trench(es) at foundation/footings is free and clear of all material and debris.

____ All excess material is stacked and protected from weather.

____ Jobsite area is clean and all excess (runoff) concrete has been removed to driveway.

____ Footing batterboards have been removed and stacked with excess material.

____ The field count of blocks has been done.

____ Field measurement of stucco is applied.

____ Field count of unused materials has been done.

Date: _____   Site Superintendent _____

Trade Contractor _____

SCOPE OF WORK
Standards and Description of Work Performance

BRICK LABOR

The Company's **Terms and Conditions** are by reference a part of all **Scope of Work** requirements.

Construction Requirements:

Generally speaking the work of Trade Contractors and their employees is expected to be performed in a good and workmanlike manner. Workmanlike quality is defined as workmanship that meets or betters those criteria indicated in applicable building codes, using materials and installation methods identified in the construction plans and this Scope of Work.

Code Requirements:

All jobs shall conform to those standards stipulated in the building code, mechanical code, plumbing code, and electrical code applicable in the local jurisdiction. All construction on The Company's jobsites shall meet or exceed NAHB Performance and Building Standards.

General Comments:

The Company considers our Trade Contractors to be experts at producing a high-quality job. But everyone on our construction team—staff, Trade Contractors, and suppliers—must recognize the importance of providing quality in both the product and service areas while on our jobsites and in the homes of our purchasers.

Since we work as a team, poor quality or service, from any of us, reflects unfavorably on all of us. An exceptional level of product quality and highly effective service can help us all to increase our business and grow.

The Company's definition of quality construction also requires that every job be completed correctly the first time. When this does not occur it costs both of us additional money, imposes on the purchaser, and hurts our reputations as quality builders. That is why, in situations where construction was not completed in a quality manner, prompt corrective action is required to remedy specific deficiencies.

In the following information the term Site Superintendent shall refer to any The Company representative with authority to perform the specified task. The term Trade Contractor shall mean the Trade Contractor's organization or any representative that is assigned the authority to perform the specified task.

General: Brickwork can be performed with a variety of methods and application techniques. More than any other trade, brickwork relies on the skill of the individual brick mason for the level of quality in the final product.

Company Rep's Initials _____
Trade Contractor's Initials _____

All of The Company's houses that require brick shall use brick veneer.

Materials: The Company will determine the type, style, and color of brick, and shall furnish the materials.

Installation: All work is to be done by trained, experienced brick masons. Bricks shall be laid level, plumb, and square. No point along the bottom of any course shall be more than 1/4 inch higher or lower than any other point within 10 feet along the bottom of the same course, or 1/2 inch in any length. Plumb lines shall be used for all installations.

Mortar should be proportioned, mixed, and applied to ensure full mortar joints. Mortar should be protected form water penetration and efflorescence and should not be allowed to freeze during construction. Mortar is workable when it spreads easily, holds the weight of the bricks, and clings without dropping off or smearing.

Masonry units should be moved to their final position while mortar is still pliable. Exposed mortar joints should be thumbprint hard before striking with a round or v-shaped joiner. Mortar should not be smeared into masonry-finished surfaces.

Brick veneer walls shall be built with an air space in the back of the brickwork to allow any penetrating water to fall to the base of the veneer where flashing and weepholes shall be provided to drain water back out of the wall. All brick veneer shall be installed over insulated sheeting. No brick shall be installed over any area without insulated sheeting or where insulation board is broken or damaged.

All head joints shall be fully filled with mortar to prevent water penetration. Wall ties shall be used 16 inches on center vertically and 32 inches on center horizontally, and fastened securely to framing members to ensure stability of brick veneer. Continuous base flashing shall be installed under the first brick course and extend up a minimum of 8 inches behind the sheathing. Flashing is to be installed around corners and lapped and sealed at joints.

"Weep holes" shall be provided ever 24 inches on center by omitting head joints in first course of outer sythe bricks above the foundation or other support.

Brick shall be thoroughly cleaned by removing large particles, thoroughly prewetting with water, applying an appropriate cleaning solution, cleaning with a brush, and rinsing thoroughly with water. Mortar on bricks is unacceptable work.

Protection of Brick: Above 40 degrees Fahrenheit (F) normal masonry procedures should be used. Walls should be covered with plastic or canvas at day's end to prevent water from entering brickwork.

Brick may be laid during temperatures of 35 to 40 degrees F if the mortar is heated to above 45 degrees F. Bricks laid during this temperature range must be protected with a covering to prevent water from entering brickwork and freezing.

Safety: The Trade Contractor must use OSHA-approved scaffolding, rigging, and other safety equipment.

Payment: Purchase orders are issued on a "best-guess" estimate. Payment will be made on the actual field measurements made at the time of the final inspection report.

_____ Company Rep's Initials
_____ Trade Contractor's Initials

Warranty: The Company believes that all work done in connection with one of our homes should be of high quality and that all Trade Contractors should stand behind the quality of their work. Therefore we require all Trade Contractors to warrant the quality of their work for a period of one (1) year from date of closing of the house. Please refer to our printed Limited Warranty booklet for specific items that are covered under Warranty as they apply to brickwork.

The Trade Contractor shall have seven (7) days in which to correct any Warranty problem. If the problem is not corrected within seven (7) days then The Company shall correct the problem and will backcharge the Trade Contractor at the rate of $25.00 per hour, with a minimum charge of $100.00 plus the cost of any materials.

Inspection Reports: The Trade Contractor and the Site Superintendent shall walk the job together and complete each section of the inspection report(s). The Trade Contractor must correct any deficiency found during the inspection and the job must be 100-percent complete before payment will be made. The Trade Contractor and the Site Superintendent must sign-off on all sections of the inspection report(s) attesting that the job is 100-percent complete and is correct per the job requirements found in this Scope of Work.

Detailed Job Requirements:

1. A new set of plans is required for each house. Plans are subject to changes and modifications. It is the responsibility of the Trade Contractor to have the new plans before beginning work. Plans should be picked up at the job trailer from the Site Superintendent. The Trade Contractor at no cost to The Company will correct any errors that occur from using an incorrect set of plans.

2. Purchase orders should be picked up from the Site Superintendent.

3. The Trade Contractor and the Site Superintendent must walk the job together and complete the pre-work section of the inspection report(s) before work may begin. Both parties must sign-off on the pre-work section of the inspection report(s).

4. Bricks shall be laid level, plumb, and square. No point along the bottom of any course shall be more than 1/4 inch higher or lower than any other point within 10 feet along the bottom of the same course, or 1/2 inch in any length. Plumb lines shall be used for all installations.

5. No brick may be laid over any area without insulation board or over insulation board that is broken or damaged.

6. Weep holes shall be provided every 24 inches on center by omitting head joints in first course of outer sythe bricks above the foundation or other support.

7. Continuous base flashing shall be installed under first brick course and extend up a minimum of 8 inches behind sheathing. Flashing is to be installed around corners and lapped and sealed at joints.

8. No brick shall be installed that has damage that would be visible in the course.

9. Joints shall be smooth and clean.

10. All brickwork shall be thoroughly cleaned, with no mortar visible on bricks for a distance of 10 feet.

Company Rep's Initials _____
Trade Contractor's Initials _____

11. All head joints shall be fully filled with mortar to prevent water penetration.

12. Wall ties shall be used 16 inches on center vertically and 32 inches on center horizontally, and fastened securely to framing members to ensure stability of brick veneer.

13. Brick shall be protected with either a plastic or canvas covering to keep water from penetrating the brick during construction.

14. All excess bricks, mortar, or other materials shall be stacked at the front of the lot on brick pallets. All mortar shall be covered with plastic to prevent moisture damage.

15. All construction debris must be removed to the dumpster or to an area designated by the Site Superintendent. The job will not be considered complete and payment will not be issued until all trash and debris have been removed from the house and/or site.

16. All excess mixed mortar must be removed to the driveway cut.

17. The Trade Contractor and Site Superintendent must count all excess material. The type and number of items must be noted on the purchase order.

18. All work must be field-measured and the field measurements must be listed on the Trade Contractor's invoice. The quantity of actual bricks used will determine payment—not the amount of brick delivered to the jobsite.

19. The Trade Contractor and Site Superintendent must walk the job together and perform a final inspection of the job. The final section of the inspection report(s) must be completed and signed-off on by both parties. The inspection report(s) must be attached to the office's copy of the purchase order and the Trade Contractor's invoice or payment will not be issued.

20. Any items found during the final inspection that need correction shall be corrected before payment will be made.

I _____ agent for _____

_____ have read and fully understand the above **Scope of Work** and
I hereby agree to perform all work in accordance with the above.

Date: _____   _____
                          Signed: Trade Contractor (or agent)

Date: _____   _____
                          For The Company

INSPECTION REPORT

BRICK LABOR

Subdivision/Lot # _____

Pre-Work Inspection: (to be completed prior to beginning work)

____   Plans have been reviewed with Site Superintendent.

____   Purchase order has been received.

____   House is ready for brick.

____   Brick lintels are installed correctly, are level and of proper length.

____   Insulation board is installed in all areas with no open areas and no broken or damaged boards.

____   Material at jobsite is ready to be used.

____   Site is clean and free of debris.

____   Windows are intact with none broken.

____   OSHA-approved scaffolding, rigging, and safety equipment is in use.

Date: _____   Site Superintendent _____

                            Trade Contractor _____

Final Inspection: (to be completed before Trade Contractor leaves jobsite)

____   Windows are intact with none broken.

____   Brick is level, plumb, and square.

____   Decorative accents are per plan.

____   Brick is completely clean with no mortar drips, smears, or runs.

____   Jobsite is clean and all debris moved to designated area.

____   Excess bricks and mortar are stacked on brick pallets. Mortar is covered in plastic.

____   Excess material is counted.

____   Field measurements have been made of actual bricks used.

____   All scaffolding and walkboards are removed with no damage to brick.

Date: _____   Site Superintendent _____

                       Trade Contractor _____

SCOPE OF WORK
Standards and Description of Work Performance

CABINETS

The Company's **Terms and Conditions** are by reference a part of all **Scope of Work** requirements.

Construction Requirements:

Generally speaking the work of Trade Contractors and their employees is expected to be performed in a good and workmanlike manner. Workmanlike quality is defined as workmanship that meets or betters those criteria indicated in applicable building codes, using materials and installation methods identified in the construction plans and this Scope of Work.

Code Requirements:

All jobs shall conform to those standards stipulated in the building code, mechanical code, plumbing code, and electrical code applicable in the local jurisdiction. All construction on The Company's jobsites shall meet or exceed NAHB Performance and Building Standards.

General Comments:

The Company considers our Trade Contractors to be experts at producing a high-quality job. But everyone on our construction team—staff, Trade Contractors, and suppliers—must recognize the importance of providing quality in both the product and service areas while on our jobsites and in the homes of our purchasers.

Since we work as a team, poor quality or service, from any of us, reflects unfavorably on all of us. An exceptional level of product quality and highly effective service can help us all to increase our business and grow.

The Company's definition of quality construction also requires that every job be completed correctly the first time. When this does not occur it costs both of us additional money, imposes on the purchaser, and hurts our reputations as quality builders. That is why, in situations where construction was not completed in a quality manner, prompt corrective action is required to remedy specific deficiencies.

In the following information the term Site Superintendent shall refer to any The Company representative with authority to perform the specified task. The term Trade Contractor shall mean the Trade Contractor's organization or any representative that is assigned the authority to perform the specified task.

General: For most buyers, the look of finished carpentry, millwork, and cabinetry defines quality construction. The look, feel, and number of cabinets are the major selling points of most homes. The first impression created by the kitchen and baths must be a pleasant one and when looked at closely must continue to be pleasant. Lack of quality workmanship in these areas "turns-off" homebuyers.

_____ Company Rep's Initials
_____ Trade Contractor's Initials

Care should be taken in the storage and handling of finish materials to avoid damage and soiling. Installed materials should be protected when necessary.

Materials: All cabinet materials, construction, type, etc., is determined by the model of the home, the location of the subdivision, and whether the cabinets are standard or upgrades.

Installation: All installations are to be performed by the Trade Contractor. All work is to be done by trained, experienced individuals. Cabinets should be installed with sufficient care to avoid damage. Cabinets should be attached with screws, not nails, to studs and other framing members. As it is attached, each cabinet should be checked front to back and across the front for level. Cabinet units may not exceed a 3/8-inch differential in 10 feet in surface alignment. A gap between the cabinet and wall that is greater than 1/4 inch is unacceptable.

Countertops should be fastened to cabinets with screws, front and back, at least every 4 feet. Care should be taken not to penetrate the countertop surface with the fasteners. Countertops should be installed within 1/4 inch of level in 8 feet and 1/4 inch of level front to back. Deck area countertop joints may not exceed a 1/16-inch gap and a maximum of 1/16-inch differential in surface alignment.

Cabinet faces more than 1/8 inch out of line, and cabinet corners more than 3/16 inch out of line are unacceptable.

Cabinet doors and drawers shall open and close with reasonable ease.

Warped cabinet faces or drawer faces shall not be installed.

Prior to the manufacture of cabinets for the house, the Trade Contractor is responsible for field measuring all cabinet areas to ensure the correct fit of the cabinets.

Warranty: The Company believes that all work done and all materials installed, in connection with one of our homes should be of high quality and that all Trade Contractors should stand behind the quality of their work and materials. Therefore we require all Trade Contractors to warrant the quality of their work for a period of one (1) year from date of closing of the house. Please refer to our printed Limited Warranty booklet for specific items that are covered under Warranty as they apply to cabinets.

The Trade Contractor shall have seven (7) days in which to correct any Warranty problem. If the problem is not corrected within seven (7) days then The Company shall correct the problem and will backcharge the Trade Contractor at the rate of $25.00 per hour, with a minimum charge of $100.00 plus the cost of any materials.

Inspection Reports: The Trade Contractor and the Site Superintendent shall walk the job together and complete each section of the inspection report(s). The Trade Contractor must correct any deficiencies found during the inspection and the job must be 100-percent complete before payment will be made. The Trade Contractor and the Site Superintendent must sign-off on all sections of the inspection report(s) attesting that the job is 100-percent complete and is correct per the job requirements found in this Scope of Work.

Company Rep's Initials _____
Trade Contractor's Initials _____

Detailed Job Requirements:

1. A new set of plans is required for each house. Plans are subject to changes and modifications. It is the responsibility of the Trade Contractor to have the new plans before beginning work. Plans should be picked up at the job trailer from the Site Superintendent. The Trade Contractor at no cost to The Company will correct any errors that occur from using an incorrect set of plans.

2. Purchase orders will be mailed to the Trade Contractor from the office.

3. Color selection sheets that detail cabinet and countertop colors may be picked up from the Site Superintendent or will be mailed from the office at least three (3) weeks prior to installation.

4. The Trade Contractor and the Site Superintendent must walk the job together and complete the pre-work section of the inspection report(s) before work may begin. The pre-work section of the inspection report(s) must be signed-off on by both parties.

5. The Trade Contractor is responsible for field measuring all cabinet areas to ensure proper sizing.

6. If plans call for marble countertops, the Trade Contractor and the marble company should compare cabinet dimensions to ensure the proper fit of cabinets and countertops.

7. Cabinets are to be built and installed to plan, with necessary cutouts for range, sinks, dishwasher, vent-a-hood, built-in microwave, etc.

8. The HVAC vent in the kick plate of the cabinet should be cut high enough to allow the HVAC grill to be above the shoe molding when the shoe molding is installed.

9. Cabinets are to be installed per the manufacturer's installation instructions, but not with less quality than is stated herein.

10. The Trade Contractor is responsible for the cabinets being level, plumb, and securely attached. A differential of more than 3/8 inch in 10 feet is unacceptable.

11. All cabinet doors and drawer fronts shall be level, plumb, and undamaged. A differential of more than 1/8 inch out of line and cabinet corners that are more than 3/16 inch out of line are unacceptable.

12. Cabinet shelves shall be level and undamaged.

13. Adjustable shelves shall rest securely on shelf holders and no shelf holder shall be missing.

14. If the cabinets have knobs, handles, or other hardware attached, the screws that attach the hardware shall be set smooth and should not damage the inside of the drawers or doors.

15. In all cabinets that include rollout shelves the roller tracks shall be attached tightly and correctly. The shelf shall roll smoothly and not tilt when fully extended. The shelf should be easy to remove for cleaning.

16. In all cabinets that contain a lazy susan, the lazy susan shall work smoothly, not tilt, and be securely anchored top and bottom.

17. All end/backsplashes must be properly fitted, snug, and secure.

18. If colored countertops are selected, colored caulking, matched as closely as possible to the color of the countertops, should be used if at all possible.

_____ Company Rep's Initials
_____ Trade Contractor's Initials

19. After cabinets are installed, all countertops should be protected with cardboard that is taped securely. Tape shall be of a type that will not damage cabinets or countertops.

20. All construction debris must be removed to the dumpster or to an area designated by the Site Superintendent. The job will not be considered complete and payment will not be issued until all trash and debris are removed from the house and/or site.

21. House is to be left clean and broom-swept.

22. The Trade Contractor and Site Superintendent must walk the job together and perform a final inspection of the job. The final section of the inspection report(s) must be completed and signed-off on by both parties. The inspection report(s) must be attached to the office's copy of the purchase order and the Trade Contractor's invoice or payment will not be issued.

23. Any items found during the final inspection that need correction shall be corrected before payment will be made.

I _____ agent for _____

_____ have read and fully understand the above **Scope of Work** and I hereby agree to perform all work in accordance with the above.

Date: _____   _____

Signed: Trade Contractor (or agent)

Date: _____   _____

For The Company

INSPECTION REPORT

CABINETS

Subdivision/Lot # _____

Pre-Work Inspection: (to be completed prior to beginning work)

____ Purchase order has been received.

____ Color selection sheet(s) has been received.

____ Cabinet areas are field-measured.

____ Windows are intact with none broken.

____ Tubs are covered and undamaged.

____ Temporary stairs (if required), temporary handrails, and safety bracing are installed.

____ House is broom-swept, clean, and free of debris

Date: _____   Site Superintendent _____

Trade Contractor _____

Final Inspection: (to be completed before Trade Contractor leaves jobsite)

____ Windows are intact with none broken.

____ Temporary handrails, safety bracing, etc., are intact.

____ Cabinets are installed per plan (correct number, size, etc.).

____ Cabinets are level and plumb.

____ Cabinets are secured correctly and tightly to wall.

____ Color of cabinets and countertops is correct.

____ Colored caulk is used if colored cabinets and countertops.

____ Cabinet doors and drawer fronts are level and square.

____ Cabinet faces and corners are level and plumb.

____ All doors and drawers open and close correctly.

____ Correct hardware is installed—square, plumb, and correctly.

____ Screws attaching hardware have not splintered wood.

____ Kick plates are installed with no gaps or holes.

____ HVAC vent is cut correctly. Shoe molding will fit below vent.

____ Cabinets are undamaged.

____ Cabinets are intact with no unfilled holes.

____ Shelves are level and undamaged.

____ The number of shelves is correct.

____   For adjustable shelves, all holders fit, none are missing, and are level.

____   For all rollout shelves, operation is smooth and shelves track tight and secure.

____   Lazy susan(s) are installed plumb, work correctly, and are secured tightly.

____   Cabinets are clean inside and out.

____   Countertops are protected with cardboard that is taped securely.

____   House is clean, broom-swept, and debris removed.

Date: _____   Site Superintendent _____

                                             Trade Contractor _____

<u>SCOPE OF WORK</u>
Standards and Description of Work Performance

CLEANING

The Company's **Terms and Conditions** are by reference a part of all **Scope of Work** requirements.

Construction Requirements:

Generally speaking the work of Trade Contractors and their employees is expected to be performed in a good and workmanlike manner. Workmanlike quality is defined as workmanship that meets or betters those criteria indicated in applicable building codes, using materials and installation methods identified in the construction plans and this Scope of Work.

Code Requirements:

All jobs shall conform to those standards stipulated in the building code, mechanical code, plumbing code, and electrical code applicable in the local jurisdiction. All construction on The Company's jobsites shall meet or exceed NAHB Performance and Building Standards.

General Comments:

The Company considers our Trade Contractors to be experts at producing a high-quality job. But everyone on our construction team—staff, Trade Contractors, and suppliers—must recognize the importance of providing quality in both the product and service areas while on our jobsites and in the homes of our purchasers.

Since we work as a team, poor quality or service, from any of us, reflects unfavorably on all of us. An exceptional level of product quality and highly effective service can help us all to increase our business and grow.

The Company's definition of quality construction also requires that every job be completed correctly the first time. When this does not occur it costs both of us additional money, imposes on the purchaser, and hurts our reputations as quality builders. That is why, in situations where construction was not completed in a quality manner, prompt corrective action is required to remedy specific deficiencies.

In the following information the term Site Superintendent shall refer to any The Company representative with authority to perform the specified task. The term Trade Contractor shall mean the Trade Contractor's organization or any representative that is assigned the authority to perform the specified task.

General: Cleaning is done in a rough and a final section with an optional touch-up after the homeowners' walk-through list is completed. Rough cleaning is done when the construction of the house is complete. Final cleaning is performed after all finish work is complete.

_____ Company Rep's Initials
_____ Trade Contractor's Initials

Rough cleaning is just what the name implies. The house shall be clean and presentable, but not in "ready-to-move-in" condition. The rough clean removes the major dirt and debris left over from construction.

Final cleaning is the final clean before the homeowners move in. This clean is done prior to the homeowners' walk-through. The final cleaning should leave the house in "ready-to-move-in" condition.

Touch-up cleaning is any cleaning that must be done after the homeowners' walk-through list is complete. The house should be returned to "move-in" condition.

The cleaning Trade Contractor has the closest contact with all areas of the house (cabinets, tubs, sinks, vinyl, carpet, etc.) and is the Trade Contractor most likely to note damage to any of these items. The Company wants to ensure that any damage done to any area of the house is repaired prior to the homeowners' walk-through. Therefore it is the responsibility of the Trade Contractor to note any damage and report it to the Site Superintendent as soon as the job is completed.

It is the intention of The Company that the homeowners be able to move into their new home and find it as close to spotless as possible. We do not expect our homeowners to have to clean their house before they can move in.

Inspection Reports: The Trade Contractor and the Site Superintendent shall walk the job together and complete each section of the inspection report(s). The Trade Contractor must correct any deficiency found during the inspection and the job must be 100-percent complete before payment will be made. The Trade Contractor and the Site Superintendent must sign-off on all sections of the inspection report(s) attesting that the job is 100-percent complete and is correct per the job requirements found in this Scope of Work.

Detailed Job Requirements:

1. The Trade Contractor and the Site Superintendent must walk the job together and complete the pre-work section of the inspection report(s) before work may begin. The pre-work section of the inspection report(s) must be signed-off on by both parties.

2. Purchase orders should be picked up from the Site Superintendent.

Rough Clean:

3. Floors should be swept and any debris removed to the dumpster or to the site designated by the Site Superintendent for trash.

4. The painters are responsible for damp wiping all overspray from windows, window frames, and trim. However, there usually will be a small amount of paint residue on windows and window frames. Windows and aluminum window frames shall have all paint residue, manufacturer's stickers, and other items removed completely (inside and outside).

5. All tubs, showers, marble tops, countertops, vinyl, etc., should have protective coverings. These coverings should stay in place until the final clean to protect the items.

6. Fireplace should be cleaned and free of debris.

7. All trim should be dusted.

Company Rep's Initials _____
Trade Contractor's Initials _____

8. Cabinets (kitchen, baths, etc.) and all cabinet drawers shall be wiped clean on the inside with no residue of sawdust, dirt, etc.

9. The Trade Contractor and Site Superintendent must walk the job together and perform a final inspection of the job. The final section of the inspection report(s) must be completed and signed-off on by both parties. The inspection report(s) must be attached to the office's copy of the purchase order and the Trade Contractor's invoice or payment will not be issued.

10. Any items found during the inspection that need correction shall be corrected before payment will be made.

Final Clean:

11. Items such as vinyl, hardwood, parquet, and oak flooring, corners on stair treads, bathtubs, showers, sinks, and countertops all should be covered and protected when the cleaning crew arrives. The cleaning crew should remove all protective coverings. Protective items that can be used again, such as bathtub covers, should be stacked in the garage to be moved to the next job. Cardboard, plastic, etc., should be removed to the dumpster or to an area designated by the Site Superintendent.

12. Using a vacuum with attachments, all walls, trim, etc., should be vacuumed to remove all dust and sanding residue. This will cut down on the amount of wet cleaning that must be done.

13. Tracks of horizontal sliding doors and windows shall be free of dirt, dust, etc.

14. Glass in windows shall be cleaned with a glass cleaner that leaves the windows clear and clean.

15. Tubs and sinks shall be cleaned with a nonabrasive cleaner, rinsed, and dried to a shine. Any chips or damages should be reported to the Site Superintendent.

16. The kitchen sink should have both the strainer and stopper in the drawer to the left of the sink or under the sink. Both of these items should be thoroughly cleaned and look new. If they cannot be cleaned the Site Superintendent is to be notified so that replacements can be received prior to the walk-through.

17. All cabinets shall be cleaned on the outside to remove streaks, construction dirt, fingerprints, etc., with a cleaner that will not damage the finish of the cabinets.

18. All Trade Contractors have been notified to place all Warranty manuals, instruction manuals, etc., in the drawer to the left of the range in the kitchen. These manuals should be returned to that drawer after cleaning.

19. All carpets shall be thoroughly vacuumed and small spots shall be cleaned. Any larger areas of dirty carpet or large stains must be report to the Site Superintendent for outside cleaning.

20. All doors, doorframes, and window ledges shall be damp-wiped.

21. All trim and base should be dusted to remove all construction dust. If any scuff marks, etc., are found on base or trim they should be removed by wiping or cleaning. If they cannot be cleaned notify the Site Superintendent.

22. Tops of showers, shower/tub combinations, handrails, and hardware shall be damped-wiped to remove all drywall dust left from sanding and finishing.

23. All light fixtures shall be dusted. Any fixture on which fingerprints, smudges, or sticky residue from labels remains must be washed rather than dusted to remove theses items.

_____ Company Rep's Initials
_____ Trade Contractor's Initials

24. All strip lighting in bathrooms shall be thoroughly cleaned to remove all fingerprints, smudges, streaks, or residue from sticky labels.

25. All handrails and pickets shall be damp-wiped to remove all dust, fingerprints, etc.

26. Hardwood, parquet, and vinyl flooring shall be thoroughly mopped with a product recommended by the manufacturer that will not damage the finish. Any damage found during cleaning to these items must be reported to the Site Superintendent immediately. Adhesive shall be completely cleaned from these surfaces.

27. All mirrors are to be clean and shiny, being sure to remove all residue of drywall dust from tops and sides.

28. All appliances shall be clean and polished on the outside and wiped clean and free of dust on the inside.

29. Unfinished basements, storage areas, and garages shall be broom-swept.

30. Finished basements shall be cleaned as if they are any other part of the house (baths, windows, etc.).

31. All thresholds shall be cleaned so they are free of dirt, mud, etc. Remove any protective coverings.

32. Faces and mantels of fireplaces shall be damp-wiped if marble or dusted if brick.

33. All ceiling fans shall be dusted and, if necessary, damp-wiped.

34. The Trade Contractor and Site Superintendent must walk the job together and perform a final inspection of the job. The final section of the inspection report(s) must be completed and signed-off on by both parties. The inspection report(s) must be attached to the office's copy of the purchase order and the Trade Contractor's invoice or payment will not be issued.

35. Any items found during the final inspection that need correction shall be corrected before payment will be made.

Optional Touch-Up:

36. After completion of the punch-out list on the homeowners' walk-through it may be necessary for the Trade Contractor to return and touch-up the house to return it to "move-in" condition. The work shall be inspected per the final section of the inspection report.

I _____ agent for _____

_____ have read and fully understand the above **Scope of Work** and I hereby agree to perform all work in accordance with the above.

Date: _____   _____
                          Signed: Trade Contractor (or agent)

Date: _____   _____
                          For The Company

INSPECTION REPORT

CLEANING (ROUGH)

Subdivision/Lot # _____

Pre-Work Inspection: (to be completed prior to beginning work)

____ House has been reviewed with Site Superintendent for any special instructions.

____ Purchase order has been received.

____ House is clean with no trash or debris.

____ Tubs, sinks, countertops, stair treads, hardwood, vinyl, etc., are all covered and protected.

____ Faceplates are in place on all switches and receptacles.

____ Cabinets are complete (hardware, shelves, etc.).

____ Appliances are installed.

____ Drywall and painting are complete.

____ All trim is installed.

____ Vinyl is installed and protected.

____ Hardwood or parquet flooring is installed and protected.

____ Doors are installed.

____ All hardware is installed.

Date: _____   Site Superintendent _____

Trade Contractor _____

Final Inspection: (to be completed before Trade Contractor leaves jobsite)

____ Floors are broom-swept and clean.

____ Windows are clean inside and out.

____ Window frames are clean inside and out.

____ Cabinets (drawers, tops, and front) are cleaned inside and out.

____ Trim is dusted.

____ House is ready for carpet and final punch-out.

____ Protection for all tubs, floors, etc., is still in place.

Date: _____   Site Superintendent _____

Trade Contractor _____

INSPECTION REPORT

CLEANING (FINAL)

Subdivision/Lot # _____

Pre-Work Inspection: (to be completed prior to beginning work)

____ Verified any special items with Site Superintendent.

____ Purchase order has been received.

____ House is clean with no trash or debris.

____ Protection for tubs, sinks, etc., is in place.

____ Faceplates are in place on all switches and receptacles.

____ Cabinets are complete with hardware, shelves, etc.

____ Appliances are installed.

____ Drywall and painting are complete.

____ All trim is installed.

____ Vinyl is installed.

____ Hardwood or parquet flooring is installed.

____ Carpet is installed.

____ Doors are installed.

____ All hardware is installed.

Date: _____   Site Superintendent _____

Trade Contractor _____

Final Inspection: (to be completed before Trade Contractor leaves jobsite)

NOTE: All areas that are to be "dusted" must be free of dust or grit when a hand is brushed across the surface.

____ All walls are dusted with no smudge marks.

____ Vinyl, hardwood, parquet flooring is mopped and cleaned with no black marks or adhesive.

____ All windowsills and trim are dusted with no smudges.

____ Handrails, stair treads, and pickets are clean and free of smudges, dirt, and dust.

____ Mirrors are clean and shiny with no drywall dust on tops or sides.

____ Sinks, showers, tubs, and toilets are clean and shiny with no drywall dust on top ledges.

____ Faucets, showerheads, and other such fixtures are clean and shiny.

____ Ductwork and return-air grills are vacuumed.

____ Windows are clean and shiny, including aluminum.

____ All hardware is clean and shiny (towel bars, toilet paper holders, etc.).

____ Cabinets are clean inside and out.

____ Countertops are clean and shiny.

____ Appliances are clean inside and out.

____ Light fixtures are clean and shiny with no fingerprints, dust, or smudges.

____ Shelving is dusted and clean.

____ Carpet is vacuumed with no spots.

____ Garage, storage rooms, and unfinished basement are clean and swept.

____ Thresholds are clean, free of dirt, mud, etc.

____ Fireplace fronts and mantels are clean and free of dust and dirt.

____ Ceiling fans are dusted or damp-wiped and clean, free of dust, smudges, etc.

____ Trash is removed to correct location.

____ Tubs and shower covers are in garage ready to be moved to next job.

____ Unfinished basement and garage are broom-swept.

____ House is in "move-in" condition.

Date: _____   Site Superintendent _____

                           Trade Contractor _____

<u>SCOPE OF WORK</u>
Standards and Description of Work Performance

CLEARING and GRADING

The Company's **Terms and Conditions** are by reference a part of all **Scope of Work** requirements.

Construction Requirements:

Generally speaking the work of Trade Contractors and their employees is expected to be performed in a good and workmanlike manner. Workmanlike quality is defined as workmanship that meets or betters those criteria indicated in applicable building codes, using materials and installation methods identified in the construction plans and this Scope of Work.

Code Requirements:

All jobs shall conform to those standards stipulated in the building code, mechanical code, plumbing code, and electrical code applicable in the local jurisdiction. All construction on The Company's jobsites shall meet or exceed NAHB Performance and Building Standards.

General Comments:

The Company considers our Trade Contractors to be experts at producing a high-quality job. But everyone on our construction team—staff, Trade Contractors, and suppliers—must recognize the importance of providing quality in both the product and service areas while on our jobsites and in the homes of our purchasers.

Since we work as a team, poor quality or service, from any of us, reflects unfavorably on all of us. An exceptional level of product quality and highly effective service can help us all to increase our business and grow.

The Company's definition of quality construction also requires that every job be completed correctly the first time. When this does not occur it costs both of us additional money, imposes on the purchaser, and hurts our reputations as quality builders. That is why, in situations where construction was not completed in a quality manner, prompt corrective action is required to remedy specific deficiencies.

In the following information the term Site Superintendent shall refer to any The Company representative with authority to perform the specified task. The term Trade Contractor shall mean the Trade Contractor's organization or any representative that is assigned the authority to perform the specified task.

General: The quality of the site preparation is extremely important. Poor quality at this stage flows all the way through construction and is reflected in the finished product.

Warranty: The Company believes that all work done in connection with one of our homes should be of high quality and that all Trade Contractors should stand behind the quality of their work.

Company Rep's Initials _____
Trade Contractor's Initials _____

Therefore we require all Trade Contractors to warrant the quality of their work for a period of one (1) year from date of closing of the house. Please refer to our printed Limited Warranty booklet for specific items that are covered under Warranty as they apply to site conditions.

The Trade Contractor shall have seven (7) days in which to correct any Warranty problem. If the problem is not corrected within seven (7) days then The Company shall correct the problem and will backcharge the Trade Contractor at the rate of $25.00 per hour, with a minimum charge of $100.00 plus the cost of any materials.

Inspection Reports: The Trade Contractor and the Site Superintendent shall walk the job together and complete each section of the inspection report(s). The Trade Contractor must correct any deficiency found during the inspection and the job must be 100-percent complete before payment will be made. The Trade Contractor and the Site Superintendent must sign-off on all sections of the inspection report(s) attesting that the job is 100-percent complete and is correct per the job requirements found in this Scope of Work.

Bad Lots: At times a lot requires an above-average amount of fill or hauling of dirt or debris. Such lots require additional work, which is treated as a change order item. Under these conditions another purchase order will be issued for the additional work. All purchase orders for additional work must be approved by the Site Superintendent and must be issued before work begins.

Detailed Job Requirements:

1. The Trade Contractor must check in with the Site Superintendent and together they must walk and review the lot.

2. Purchase orders should be picked up from the Site Superintendent.

3. The Trade Contractor and the Site Superintendent must walk the job together and complete the pre-work section of the inspection report(s) before work begins. The pre-work section of the inspection report(s) must be signed-off on by both parties.

4. The Trade Contractor and Site Superintendent must walk the job together and perform a final inspection of the job. The final section of the inspection report(s) must be completed and signed-off on by both parties. The inspection report(s) must be attached to the office's copy of the purchase order and the Trade Contractor's invoice or payment will not be issued.

5. Any items found during the final inspection that need correction shall be corrected before payment will be made.

6. No debris of any kind may be buried on any lot.

NOTE: Clearing and Grading is divided into 3 sections—Grading 1, Grading 2, and Grading 3.

Grading 1:

7. The Trade Contractor and the Site Superintendent must complete the pre-work section of the Grading 1inspection report before work may begin.

8. Trees, stumps, and any debris left over from development shall be removed.

9. Rough grading and leveling of the lot shall leave no major hills, valleys, holes, or excess trees.

_____ Company Rep's Initials
_____ Trade Contractor's Initials

10. The Trade Contractor shall prepare a clear and level pad area.

11. The Trade Contractor shall cut the driveway.

12. The Trade Contractor and the Site Superintendent must complete the final inspection section of the inspection report for Grading 1 before the Trade Contractor leaves the jobsite.

Grading 2:

13. The Trade Contractor and the Site Superintendent must complete the pre-work section of the Grading 2 inspection report before work may begin.

14. The slab shall be backfilled, taking care not to damage rough plumbing, foundation, footings, etc.

15. The Trade Contractor shall grade front- and backyards to create swales and berms, as required, to ensure correct drainage of the lot.

16. The Trade Contractor shall divert water to either the front of the house (to the street) or to natural creeks, etc. No water is to be diverted onto neighboring lots.

17. Excess dirt shall be removed to a location determined by the Site Superintendent.

18. The Trade Contractor and the Site Superintendent must complete the final inspection section of the inspection report for Grading 2 before the Trade Contractor leaves the jobsite.

Grading 3:

19. The Trade Contractor and the Site Superintendent must complete the pre-work section of the Grading 3 inspection report before work may begin.

20. Final grading of the lot shall leave it ready for landscaping.

21. Final grading around the house shall be to a level a minimum of 8 inches below the top of the foundation. Inspections require a full 8 inches of exposed foundation.

22. Final grading shall repair the original swales and berms if any damage has occurred, and shall ensure that the direction of the water flow is still correct.

23. The Trade Contractor and the Site Superintendent must complete the final inspection section of the inspection report for Grading 2 before the Trade Contractor leaves the jobsite.

I _____ agent for _____

_____ have read and fully understand the above **Scope of Work** and I hereby agree to perform all work in accordance with the above.

Date: _____   _____
                            Signed: Trade Contractor (or agent)

Date: _____   _____
                            For The Company

INSPECTION REPORT

CLEARING AND GRADING (GRADING 1)

Subdivision/Lot # _____

Pre-Work Inspection: (to be completed prior to beginning work)

____   Purchase order has been received.

____   House area is staked and lot pins are in place.

____   Area to be cleared is clearly marked and the Site Superintendent and the Trade Contractor have walked the area together.

____   Any natural undisturbed area is marked and understood.

____   Direction of water drainage has been determined and is understood.

____   Area to place trees, stumps, and development debris has been designated by the Site Superintendent and the Trade Contractor understands the location and how the debris is to be collected, etc.

Date: _____   Site Superintendent _____

    Trade Contractor _____

Final Inspection: (to be completed before Trade Contractor leaves jobsite)

____   All debris has been removed from the lot.

____   Lot is roughly level with no holes, hills, valleys, etc.

____   Pad has been cut and measured to determine correctness of size.

____   House pad is correctly located.

____   Driveway is cut and ready for stone.

____   Undisturbed natural area is still undisturbed.

Date: _____   Site Superintendent _____

    Trade Contractor _____

INSPECTION REPORT

CLEARING AND GRADING (GRADING 2)

Subdivision/Lot # _____

Pre-Work Inspection: (to be completed prior to beginning work)

____   Purchase order has been received.

____   Lot is clear of any construction debris.

____   Foundation is complete.

____   Rough mechanicals are installed in slab area.

Date: _____   Site Superintendent _____

                          Trade Contractor _____

Final Inspection: (to be completed before Trade Contractor leaves jobsite)

____   Slab is backfilled. All rough mechanicals are intact and undamaged.

____   All foundation drains are clear of dirt and extend 6 inches above top of footings.

____   Front and backyard are graded.

____   Swales and berms are established.

____   Water flow direction is correct.

____   All excess dirt has been removed.

____   Lot is clear and free of building debris.

Date: _____   Site Superintendent _____

                          Trade Contractor _____

INSPECTION REPORT

CLEARING AND GRADING (GRADING 3)

Subdivision/Lot # _____

Pre-Work Inspection: (to be completed prior to beginning work)

____   Purchase order has been received.

____   Lot is clear and free of construction debris.

____   Any repairs required for swales and berms are understood.

Date: _____   Site Superintendent _____

                               Trade Contractor _____

Final Inspection: (to be completed before Trade Contractor leaves jobsite)

____   Swales and berms are correct. Water direction is correct.

____   Final grading of the lot is complete. Lot is ready for landscaping,

____   Eight (8) inches of foundation clear around entire house.

____   Foundation drains are visible and clear.

Date: _____   Site Superintendent _____

                               Trade Contractor_____

SCOPE OF WORK
Standards and Description of Work Performance

DECK and PORCH LABOR

The Company's **Terms and Conditions** are by reference a part of all **Scope of Work** requirements.

Construction Requirements:

Generally speaking the work of Trade Contractors and their employees is expected to be performed in a good and workmanlike manner. Workmanlike quality is defined as workmanship that meets or betters those criteria indicated in applicable building codes, using materials and installation methods identified in the construction plans and this Scope of Work.

Code Requirements:

All jobs shall conform to those standards stipulated in the building code, mechanical code, plumbing code, and electrical code applicable in the local jurisdiction. All construction on The Company's jobsites shall meet or exceed NAHB Performance and Building Standards.

General Comments:

The Company considers our Trade Contractors to be experts at producing a high-quality job. But everyone on our construction team—staff, Trade Contractors, and suppliers—must recognize the importance of providing quality in both the product and service areas while on our jobsites and in the homes of our purchasers.

Since we work as a team, poor quality or service, from any of us, reflects unfavorably on all of us. An exceptional level of product quality and highly effective service can help us all to increase our business and grow.

The Company's definition of quality construction also requires that every job be completed correctly the first time. When this does not occur it costs both of us additional money, imposes on the purchaser, and hurts our reputations as quality builders. That is why, in situations where construction was not completed in a quality manner, prompt corrective action is required to remedy specific deficiencies.

In the following information the term Site Superintendent shall refer to any The Company representative with authority to perform the specified task. The term Trade Contractor shall mean the Trade Contractor's organization or any representative that is assigned the authority to perform the specified task.

General: All decks and porches shall meet or exceed all county, state, and federal building codes for safety.

Installation: All work shall be performed by trained, experienced personnel. The Trade Contractor shall follow all OSHA safety requirements including the use of scaffolding, handrails on scaffolding,

Company Rep's Initials _____
Trade Contractor's Initials _____

etc. All work shall meet or exceed local, state, or national safety standards for decks, deck landings, and deck stairs.

Warranty: The Company's printed Limited Warranty booklet excludes porches and decks from the Warranty coverage being administrated by our chosen national warranty administration company. However, The Company believes that all work done in connection with one of our homes should be of quality work and all Trade Contractors should stand behind the quality of their work. Therefore, we require all Trade Contractors who install decks and porches to warrant their work for a period of one (1) year from the closing of the house.

The Warranty shall be: Any deck, rail, stair, porch, landing, or other item constructed by the Trade Contractor shall meet or exceed the safety requirements of local, state, or federal authorities. The Trade Contractor is responsible for the repair or correction of any item that fails any safety codes. Any movement in stairs, deck rails, and pickets that exceeds the allowable limits of state or local codes shall be corrected by the Trade Contractor. Any wood that splits due to incorrect installation shall be replaced by the Trade Contractor. Any work that is found to be of subquality shall be repaired or replaced by the Trade Contractor.

The Trade Contractor shall have seven (7) days in which to correct any Warranty problem. If the problem is not corrected within seven (7) days then The Company shall correct the problem and will backcharge the Trade Contractor at the rate of $25.00 per hour, with a minimum charge of $100.00 plus the cost of any materials.

Inspection Reports: The Trade Contractor and the Site Superintendent shall walk the job together and complete each section of the inspection report(s). The Trade Contractor must correct any deficiency found during the inspection and the job must be 100-percent complete before payment will be made. The Trade Contractor and the Site Superintendent must sign-off on all sections of the inspection report(s) attesting that the job is 100-percent complete and is correct per the job requirements found in this Scope of Work.

Detailed Job Requirements:

1. A new set of plans is required for each house. Plans are subject to changes and modifications. It is the responsibility of the Trade Contractor to have the new plans before beginning work. Plans should be picked up at the job trailer from the Site Superintendent. The Trade Contractor at no cost to The Company will correct any errors that occur from using an incorrect set of plans.

2. The Trade Contractor and the Site Superintendent must walk the job together and complete the pre-work section of the inspection report(s) before work may begin. The pre-work section of the inspection report(s) must be signed-off on by both parties.

3. Purchase orders should be picked up from the Site Superintendent.

4. The Company will furnish all materials.

5. Decks shall be built to plan.

_____ Company Rep's Initials
_____ Trade Contractor's Initials

6. Local, state, or national safety codes shall take precedence over any item below should such items be in conflict.

7. Decking material that has excessive splits (light visible through splits or splits that penetrate more than half of the board) shall not be used.

8. Decking material that is warped, twisted, or otherwise more than 1/4 inch out of level shall not be used.

9. All decking will be 5/4-board bullnosed. The spacing on opposite sides of individual deck boards shall not differ in average width by more than 3/16 inch.

10. All nails will be ringshank nails. No nailheads shall protrude from the wood decking.

11. Decks or porches will be bolted to the house using 1/2-inch bolts. No lags are acceptable.

12. Jacks will be shackle-bolted and the shackle set in concrete.

13. All corners of all handrails will terminate with a 4x4 deck post including handrails at stairs, landings, and main deck.

14. Railings on wood decks shall not contain slivers longer than 1/8 inch in exposed areas.

15. No point on the deck surface shall be more than 1/2 inch higher or lower than any other deck surface point within 10 feet on a line parallel to the house, or proportional multiples of the preceding dimensions.

16. Landings shall be installed as required by safety codes.

17. All handrails should be without miter cuts.

18. All handrails will be tight and secure with no give, shaking, or movement.

19. All 2x2 pickets shall be nailed securely and without splitting.

20. The Trade Contractor shall keep waste material to a minimum. Any unnecessary waste or damage to materials will be at the expense of the Trade Contractor.

21. Any excess material is to be counted, noted on the purchase order, and stacked in garage.

22. All unusable material (due to splits, twisted, or warped boards, etc.) shall be counted and noted as to type and quantity on the purchase order.

23. All construction debris must be removed to the dumpster or to an area designated by the Site Superintendent. The job will not be considered complete and payment will not be issued until all trash and debris have been removed from the site.

24. The Trade Contractor and Site Superintendent must perform a final inspection and they both must sign-off on the final section of the inspection report. The inspection report must be attached to the purchase order and the Trade Contractor's invoice or payment will not be made.

25. Any items found during the final inspection that need correction shall be corrected before payment will be made.

Company Rep's Initials _____
Trade Contractor's Initials _____

I _____ agent for _____

_____ have read and fully understand the above **Scope of Work** and
I hereby agree to perform all work in accordance with the above.

Date: _____   _____
                          Signed: Trade Contractor (or agent)

Date: _____   _____
                          For The Company

INSPECTION REPORT

DECK AND PORCH LABOR

Subdivision/Lot # _____

Initial Inspection: (to be completed prior to beginning work)

____ Deck and/or porch plans are correct.

____ Purchase order has been picked up.

____ Site is clean and free of debris.

____ House is sided and ready for deck/porch.

____ Material is on site and ready to be used.

____ Windows are intact with none broken.

____ Electricity is available.

Date: _____   Site Superintendent _____

                            Trade Contractor _____

Final Inspection: (to be completed before Trade Contractor leaves jobsite)

____ Deck is built per plan.

____ Porch is built per plan.

____ Deck/porch is level.

____ Decking is 5/4-board bullnosed. Spacing is correct.

____ Deck/porch is firmly bolted to house with no give or sway.

____ Deck posts are 4x4 and installed plumb.

____ Jacks are shackle-bolted and set in concrete.

____ All corners of all handrails terminate with a 4x4 deck post.

____ Handrails are secure with no give, sway, or movement.

____ Pickets are secure and correctly nailed.

____ Step run and rise are correct.

____ Landings are per plan if required.

____ Jobsite is clean. Construction debris and trash have been removed to correct area.

____ Excess material is stacked in garage, field counted, and noted on purchase order.

____ Unusable material is counted and noted on purchase order.

Date: _____   Site Superintendent _____

                            Trade Contractor _____

SCOPE OF WORK
Standards and Description of Work Performance

DRIVES and WALKWAYS

The Company's **Terms and Conditions** are by reference a part of all **Scope of Work** requirements.

Construction Requirements:

Generally speaking the work of Trade Contractors and their employees is expected to be performed in a good and workmanlike manner. Workmanlike quality is defined as workmanship that meets or betters those criteria indicated in applicable building codes, using materials and installation methods identified in the construction plans and this Scope of Work.

Code Requirements:

All jobs shall conform to those standards stipulated in the building code, mechanical code, plumbing code, and electrical code applicable in the local jurisdiction. All construction on The Company's jobsites shall meet or exceed NAHB Performance and Building Standards.

General Comments:

The Company considers our Trade Contractors to be experts at producing a high-quality job. But everyone on our construction team—staff, Trade Contractors, and suppliers—must recognize the importance of providing quality in both the product and service areas while on our jobsites and in the homes of our purchasers.

Since we work as a team, poor quality or service, from any of us, reflects unfavorably on all of us. An exceptional level of product quality and highly effective service can help us all to increase our business and grow.

The Company's definition of quality construction also requires that every job be completed correctly the first time. When this does not occur it costs both of us additional money, imposes on the purchaser, and hurts our reputations as quality builders. That is why, in situations where construction was not completed in a quality manner, prompt corrective action is required to remedy specific deficiencies.

In the following information the term Site Superintendent shall refer to any The Company representative with authority to perform the specified task. The term Trade Contractor shall mean the Trade Contractor's organization or any representative that is assigned the authority to perform the specified task.

General: Concrete is subject to various phenomena including applied chemicals and natural elements that can deteriorate the surface. In some cases, however, surface defects such as flaking, scaling, or spallling may be caused by improper finishing. Excessive powdering or chalking may occur due to improper troweling, excess water, or when uncured concrete freezes. It is important

_____ Company Rep's Initials
_____ Trade Contractor's Initials

that proper precautions and correct techniques be utilized in the handling and finishing of concrete in the field to ensure a quality job.

Concrete shall maintain its integrity. Poured concrete drives and walks should not crack in excess of 1/8 inch in width or vertical displacement. Concrete shall not pit, scale, or spall to the extent that the aggregate is exposed under normal weathering and use.

Outdoor drives, walks, stoops, steps, and patios should be designed so that water drains off.

Installation: All work is to be done by trained, experienced individuals. Forms should be tight and well braced. They should be moistened in hot weather to prevent water extraction. Snow and ice must be removed from forms prior to pouring.

Concrete should be placed at an appropriate rate so that it can be spread, straightened, and darbied properly. Techniques for handling and placing concrete should ensure that it remains uniform within each batch and from batch to batch. Concrete should not be allowed to run or be worked over long distances and should not be allowed to drop more than 4 feet.

Finishing: Concrete should be placed in a timely and continuous manner. Excess concrete should be screeded off. To avoid walking onto a screeded area, screed stakes should be removed as the work progresses.

Screeding should be followed immediately by darbying once to bring mortar to the true surface grade. Darbying should embed coarse aggregate and eliminate voids and ridges left by screeding.

When the concrete has lost its sheen and begins to stiffen, control joints may be cut and the area may be edged. The concrete surface then should be floated to remove high and low spots, surface imperfections left by edging, or jointing, and to consolidate the mortar at the surface.

The surface then should be immediately brushed lightly to a nonslip surface.

Curing: To ensure maximum strength, concrete should be protected from rapid drying by covering with polyethylene. Forms should be left in place as long as practical. The surface should be kept uniformly wet or moist through the period of curing. In cold weather, with air temperatures below 40 degrees Fahrenheit (F), the water and aggregate should be heated so that the mixed concrete is placed at a temperature between 50 and 70 degrees F. No drive may be poured on frozen ground. The temperature must be above freezing and **rising** before pouring. If there is a possibility of frost then 2-percent calcium should be added to the concrete.

Payment: The price on the purchase order is a "best-guess" estimate of the size of drives and walks. Field measurements by the Trade Contractor and Site Superintendent are necessary to determine the payment amount. Payment will **not** be based on the yards of concrete delivered to the jobsite.

Warranty: Our printed Limited Warranty booklet excludes walks and drives from the Warranty coverage being administrated by our chosen national warranty administration company. However, The Company believes that all work done in connection with one of our homes should be of quality work and that all Trade Contractors should stand behind the quality of their work. Therefore, we

Company Rep's Initials _____
Trade Contractor's Initials _____

require all Trade Contractors who install walks, drives, and patios to warrant their work for a period of one (1) year from date of closing of the house.

Warranty Shall Be: No drive, walkway, or patio shall scale, spall, or crack in excess of 1/4 inch or have any vertical displacement in excess of 1/8 inch. No aggregate shall be exposed under normal weathering conditions. No water shall stand on any drive, patio, or walkway.

The Trade Contractor shall have seven (7) days in which to correct any Warranty problem. If the problem is not corrected within seven (7) days then The Company shall correct the problem and will backcharge the Trade Contractor at the rate of $25.00 per hour, with a minimum charge of $100.00 plus the cost of any materials.

Inspection Reports: The Trade Contractor and the Site Superintendent shall walk the job together and complete each section of the inspection report(s). The Trade Contractor must correct any deficiency found during the inspection and the job must be 100-percent complete before payment will be made. The Trade Contractor and the Site Superintendent must sign-off on all sections of the inspection report(s) attesting that the job is 100-percent complete and is correct per the job requirements found in this Scope of Work.

Detailed Job Requirements:

1. The Trade Contractor and the Site Superintendent must walk the job together and complete the pre-work section of the inspection report(s) before work may begin.

2. Purchase orders should be picked up from the Site Superintendent.

3. The Trade Contractor and the Site Superintendent shall set forms together to ensure the accuracy of drive and walks.

4. The Trade Contractor shall fine-grade for drive and walks.

5. Concrete drives and walks shall be poured and finished to a minimum of 4 inches in depth. Concrete in all walks and drives shall be 2,500 psi.

6. Expansion joints are required every 12 feet in drives and walks. Proportional expansion joints shall be installed in a specific grid when the width is greater than 12 feet.

7. The Trade Contract shall broom-finish all walks and drives.

8. Walks/drives shall be constructed and finished to ensure proper water runoff with no standing water.

9. No aggregate shall be exposed in any portion of the drive or walks.

10. If for any reason the patio and/or porch was not poured at the same time as the slab, the patio/porch shall be poured using expansion joints at connection areas. The patio/porch slabs shall be 1-1/2 inch lower than the finished floor of the base building and slope 1/4 inch per foot to the outside edge to ensure proper water runoff. The patio/porch shall be poured to a minimum of 4 inches in depth.

_____ Company Rep's Initials
_____ Trade Contractor's Initials

11. All runoff concrete and construction debris must be removed to the dumpster or to an area designated by the Site Superintendent. The job will not be considered complete and payment will not be issued until all trash and debris have been removed from the site.

12. All formboards are to be left on drives and walks to ensure full curing. They will be removed just prior to landscaping by the landscaper.

13. All drives and walks (and patio/porch if poured at the same time) shall be field-measured by the Site Superintendent and Trade Contractor. Actual square footage shall be included on the purchase order. Both the Site Superintendent and the Trade Contractor must sign the purchase order attesting to the field measurements.

14. The Trade Contractor and Site Superintendent must walk the job together and perform a final inspection of the job. The final section of the inspection report(s) must be completed and signed-off on by both parties. The inspection report(s) must be attached to the office's copy of the purchase order and the Trade Contractor's invoice or payment will not be issued.

15. Any items found during the final inspection that need correction shall be corrected before payment will be made.

I _____ agent for _____

_____ have read and fully understand the above **Scope of Work** and I hereby agree to perform all work in accordance with the above.

Date: _____

_____
Signed: Trade Contractor (or agent)

Date: _____

_____
For The Company

INSPECTION REPORT

DRIVES and WALKWAYS

Subdivision/Lot # _____

Pre-Work Inspection: (to be completed prior to beginning work)

____ Plans have been reviewed with Site Superintendent.

____ Purchase order has been picked up.

____ Work area is clean and free of debris.

____ Drive and walks are staked.

____ Dimensions are correct.

____ All materials are on site and ready to be used.

____ Clear access and are stable base is in place for concrete trucks.

Date: _____   Site Superintendent _____

Trade Contractor _____

Final Inspection: (to be completed before Trade Contractor leaves jobsite)

____ Drive connection to garage is such that no water may enter garage.

____ Drive, walkways, and patio/porch have been checked to ensure proper water runoff.

____ No aggregate is exposed.

____ No cracks or displacements are visible.

____ Expansion joints are installed properly.

____ Dimensions are correct.

____ Surface is broom-finished.

____ Any concrete runoff has been removed.

____ Jobsite is clean and free of debris.

____ Drive barricade board is installed.

____ Field count is completed and noted on purchase order.

Date: _____   Site Superintendent _____

Trade Contractor _____

<u>SCOPE OF WORK</u>
Standards and Description of Work Performance

DRYWALL

The Company's **<u>Terms and Conditions</u>** are by reference a part of all **<u>Scope of Work</u>** requirements.

Construction Requirements:

Generally speaking the work of Trade Contractors and their employees is expected to be performed in a good and workmanlike manner. Workmanlike quality is defined as workmanship that meets or betters those criteria indicated in applicable building codes, using materials and installation methods identified in the construction plans and this Scope of Work.

Code Requirements:

All jobs shall conform to those standards stipulated in the building code, mechanical code, plumbing code, and electrical code applicable in the local jurisdiction. All construction on The Company's jobsites shall meet or exceed NAHB Performance and Building Standards.

General Comments:

The Company considers our Trade Contractors to be experts at producing a high-quality job. But everyone on our construction team—staff, Trade Contractors, and suppliers—must recognize the importance of providing quality in both the product and service areas while on our jobsites and in the homes of our purchasers.

Since we work as a team, poor quality or service, from any of us, reflects unfavorably on all of us. An exceptional level of product quality and highly effective service can help us all to increase our business and grow.

The Company's definition of quality construction also requires that every job be completed correctly the first time. When this does not occur it costs both of us additional money, imposes on the purchaser, and hurts our reputations as quality builders. That is why, in situations where construction was not completed in a quality manner, prompt corrective action is required to remedy specific deficiencies.

In the following information the term Site Superintendent shall refer to any The Company representative with authority to perform the specified task. The term Trade Contractor shall mean the Trade Contractor's organization or any representative that is assigned the authority to perform the specified task.

<u>General:</u> Drywall panels should be cut to fit easily into place, with no gaps greater than 1/2 inch. Drywall should be securely fastened using fasteners of a type recommended for the intended application. Fasteners should be driven squarely and placed at least 3/8 inch from the edges and ends. Heads should be seated no deeper than 1/32 inch below board surface and **should not break the facepaper.**

Company Rep's Initials _____
Trade Contractor's Initials _____

The finished drywall should present a smooth, unblemished, homogeneous appearance with inconspicuous joining between boards and no visible fasteners. There should be no areas of raised fibers on the facepaper due to over-sanding. Clearly visible nail pops, seam lines and cracks are considered unacceptable through the first-year Warranty period and are absolutely unacceptable at the completion of the job.

Materials: Drywall panel delivery should coincide with installation as closely as possible. The panels should be stored inside under cover on a flat surface. They may be stored vertically for a **short** period of time and should remain wrapped (if delivered wrapped) until ready to use. If they must be stored vertically they should rest only against a load-bearing wall.

Joint compound and tape should conform to ASTM C475 Treatment Materials for Gypsum Wallboard. Premixed compounds shall be used and kept from freezing. Edge and corner trim should be protected from damage before installation.

Corner beads shall be used on all corners. No metal corners are permitted. The appropriate beads shall be used on all vault and tray ceilings. Out-of-alignment vault or tray lines are unacceptable and must be repaired at the Trade Contractor's expense.

Installation: All work is to be done by trained, experienced individuals. Nail popping, visible seam lines, and cracking can be minimized with proper panel installation. Marking panel stud locations and the use of correct nails and screws can help eliminate nail pops. Applying sufficient pressure against panels to ensure secure nail attachment also can help limit nail pops.

Seam lines and cracking can be minimized by avoiding large gaps during installation and by installing panels in a manner that avoids conspicuous butt-end joints. When butt-end joints must occur, they should be staggered and as far from the center of walls and ceilings as possible. Whenever possible, panel ends and edges that are parallel to supporting members should fall on those members.

Measurements should be taken accurately at the point of installation to allow for irregularities in framing. Cut edges should be smoothed to fit accurately. Ceiling panels should be placed first and cut so as to fit easily into place without forcing. Tapered panel edges should butt tapered edges and square job-cut or mill-cut ends should butt other square-cut ends.

Fasteners should be applied starting in the middle of the panel and moving toward the outside. Nails or screws should be seated squarely while the board is held in firm contact with the framing support. Where adhesive is used, bonding surfaces should be free of dirt, grease, oil, or other foreign materials.

All beads must be installed per the manufacturer's instructions. All corners, ceiling lines, etc., must be straight with no hairline cracks.

Finishing: Three coats of joint compound are required: an embedding coat to bond the tape and two finishing coats over the tape. Each coat must be thoroughly dry before the next is applied to ensure that the surface incurs maximum shrinkage and can be readily sanded. For the final coat, sufficient lighting must be utilized to ensure a quality finish.

Turnkey Job: If the Trade Contractor furnishes material and labor The Company will issue a joint check to the Trade Contractor and the drywall supplier

_____ Company Rep's Initials
_____ Trade Contractor's Initials

Warranty: The finished drywall should present a smooth, unblemished, homogeneous appearance with inconspicuous joining between boards and no visible fasteners. There should be no areas of raised fibers on the facepaper due to over-sanding. Clearly visible nail pops, seam lines, and cracks are considered unacceptable through the first-year Warranty period. Refer to The Company's Limited Warranty booklet for a full description of warrantable items.

The Trade Contractor shall have seven (7) days in which to correct any Warranty problem. If the problem is not corrected within seven (7) days then The Company shall correct the problem and will backcharge the Trade Contractor at the rate of $25.00 per hour, with a minimum charge of $100.00 plus the cost of any materials.

Inspection Reports: The Trade Contractor and the Site Superintendent shall walk the job together and complete each section of the inspection report(s). The Trade Contractor must correct any deficiency found during the inspection and the job must be 100-percent complete before payment will be made. The Trade Contractor and the Site Superintendent must sign-off on all sections of the inspection report(s) attesting that the job is 100-percent complete and is correct per the job requirements found in this Scope of Work.

Detailed Job Requirements:

1. The Trade Contractor and the Site Superintendent must walk the job together and complete the pre-work section of the inspection report(s) before work may begin. The pre-work section of the inspection report(s) must be signed-off on by both parties.

2. Purchase orders should be picked up from the Site Superintendent.

3. Drywall and bead material will be furnished by The Company and will be on site. The Trade Contractor will furnish adhesive, mud, paper tapes, and any other necessary supplies.

4. Drywall is to be installed either vertically or horizontally depending on the most efficient use of the material.

5. Trade Contractor will keep waste to a minimum.

6. Ceilings will be stippled unless notified otherwise. Stipple shall be uniform with no thin or missed spots. Stipple must be uniform from room to room.

7. All corners, lines of tray, and vault ceilings shall have beads and be installed per the manufacturer's instructions. No metal corners are acceptable. If metal corners are found, it will be the responsibility of the Trade Contractor to remove the metal corners and replace with bead corners.

8. Drywall should be cut accurately and should fit together with a maximum 1/2-inch gap between sheets.

9. No cut-on-site edges shall be used in vault ceilings. Factory-cut edges must be used at the peak of the vault.

Company Rep's Initials _____
Trade Contractor's Initials _____

10. Drywall should be firmly attached to all framing using nails, screws, and adhesive. There should be no give, bows, or warps in any board.

11. No floating of drywall is allowed. In rare instances the Trade Contractor will have to install deadwood to ensure a tight and smooth fit.

12. Heads of fasteners should be seated no deeper than 1/32 inch below board surface and **should not break the facepaper**. Fasteners shall be installed every 16 inches on center with the use of adhesive.

13. All drywall edges shall be smooth and clean. Drywall shall not be cut with anything that leaves a ragged edge.

14. Hairline cracks in corners are unacceptable.

15. There should be no areas of raised fibers due to over-sanding.

16. All electrical boxes are to be free and clean of any drywall debris or mud before the Trade Contractor leaves the job.

17. Drywall installed around tubs and showers shall fit snugly with no gaps larger than 1/4 inch.

18. Drywall around the fireplace box, any door, window, or electrical box shall fit level and snug against jacks, fireplace box, or electrical boxes. If drywall is broken around electrical boxes, it shall be cut and patched with drywall to fit snugly around the box.

19. The Trade Contractor and its employees are not to stand on tubs, sinks, or countertops, or place any materials on them. Any fasteners dropped on tubs and sinks shall be picked up so as not to damage the finish of tubs and sinks.

20. The Trade Contractor shall not leave any holes in storage rooms, around heating units, etc. All areas that receive drywall shall be taped, mudded, and sanded.

21. Trade Contractor is responsible for cleaning up all drywall mud from floors.

22. Trade Contractor is responsible for removing all scraps and debris and placing same in the dumpster or designated trash site. The house shall be broom-swept before job is considered complete.

23. After primer coat of interior paint is applied to walls, Trade Contractor is responsible for touch-up on walls and ceilings at no additional charge.

24. Trade Contractor is responsible for a final touch-up, as required, after the homeowners' walk-through.

25. The Trade Contractor shall furnish heaters as required. The Trade Contractor is not required to leave his or her own heaters overnight unless they are returning the next day. If the job requires heat overnight notify the Site Superintendent.

26. All excess material should be stacked in garage.

_____ Company Rep's Initials
_____ Trade Contractor's Initials

27. All excess material is to be counted and noted on the purchase order.

28. Trade Contractor and Site Superintendent must perform a final inspection and they both must sign-off on the final section of the inspection report. The inspection report must be attached to the purchase order and the Trade Contractor's invoice or payment will not be made.

29. Final inspection of all drywall shall be done in sunlight and also in normal house lighting. Drywall cannot display any defects that can be readily seen at a distance of 6 feet under these conditions.

30. Any items found during the final inspection that need correction shall be corrected before payment will be made.

I _____ agent for _____

_____ have read and fully understand the above **Scope of Work** and I hereby agree to perform all work in accordance with the above.

Date: _____   _____
                            Signed: Trade Contractor (or agent)

Date: _____   _____
                            For The Company

INSPECTION REPORT

DRYWALL

Initial Inspection: (to be completed prior to beginning work)

____   House plans have been reviewed.

____   Purchase order has been picked up.

____   House is clean and free of debris.

____   Material is stocked and ready to be used.

____   Windows are intact—none broken.

____   Temporary stairs, temporary handrails, and safety bracing are in place.

____   Sinks, tubs, and countertops are covered and protected.

____   Cabinets are installed and undamaged.

____   Ducts are covered.

____   All house details are framed (access holes, stairways, special ceilings, scuttle holes, etc.).

____   Framing items are correct—no bowed studs, no nails protruding into drywall area.

____   Fireplace is set and is level and plumb.

Date: _____   Site Superintendent _____

                                   Trade Contractor _____

Final Inspection: (to be completed before Trade Contractor leaves jobsite)

____   Drywall is tightly fitted to walls.

____   Drywall does not float in any area and is secure at all openings (closets, windows, doors, fireplace, etc.).

____   Nails and screws are 16 inches on center.

____   All fasteners are mudded, taped, and sanded.

____   Joints are smooth and sanded.

____   There are no gaps, spaces, or broken pieces of drywall at electrical boxes.

____   Electrical boxes are free of mud and debris.

____   There are no gaps larger than 1/4 inch around tubs, showers, fireboxes, windows, and doors.

____   There are no hairline cracks in any corner, vault, tray, or ceiling line.

____   There are no undrywalled holes left in storage rooms, garages, etc.

____   **All** drywall joints are taped, mudded, and sanded, including storage rooms, etc.

____   Paper is smooth, not over-sanded.

____   Walls are flat and smooth, with no visible bumps in joints due to rough edges.

____   Center line of vault is straight and crisp.

____   Lines of tray ceiling are straight and crisp.

____ All corners are taped and smooth with no metal corners.

____ Temporary handrails and bracing are intact.

____ There are no broken windows.

____ Tubs, showers, sinks, and countertops are covered and undamaged.

____ Drywall mud has been removed from floor. Floor is broom-swept.

____ All debris has been removed to designated site.

____ Excess materials are counted and noted on purchase order.

Date: _____   Site Superintendent _____

                               Trade Contractor _____

SCOPE OF WORK
Standards and Description of Work Performance

ELECTRIC

The Company's **Terms and Conditions** are by reference a part of all **Scope of Work** requirements.

Construction Requirements:

Generally speaking the work of Trade Contractors and their employees is expected to be performed in a good and workmanlike manner. Workmanlike quality is defined as workmanship that meets or betters those criteria indicated in applicable building codes, using materials and installation methods identified in the construction plans and this Scope of Work.

Code Requirements:

All jobs shall conform to those standards stipulated in the building code, mechanical code, plumbing code, and electrical code applicable in the local jurisdiction. All construction on The Company's jobsites shall meet or exceed NAHB Performance and Building Standards.

General Comments:

The Company considers our Trade Contractors to be experts at producing a high-quality job. But everyone on our construction team—staff, Trade Contractors, and suppliers—must recognize the importance of providing quality in both the product and service areas while on our jobsites and in the homes of our purchasers.

Since we work as a team, poor quality or service, from any of us, reflects unfavorably on all of us. An exceptional level of product quality and highly effective service can help us all to increase our business and grow.

The Company's definition of quality construction also requires that every job be completed correctly the first time. When this does not occur it costs both of us additional money, imposes on the purchaser, and hurts our reputations as quality builders. That is why, in situations where construction was not completed in a quality manner, prompt corrective action is required to remedy specific deficiencies.

In the following information the term Site Superintendent shall refer to any The Company representative with authority to perform the specified task. The term Trade Contractor shall mean the Trade Contractor's organization or any representative that is assigned the authority to perform the specified task.

General: The electrical system should be complete and functioning, system-tested, and ready for operation. The wiring equipment, materials, and methods shall be in full compliance with the latest edition of the National Electrical Code and be acceptable to the locality and building code

_____ Company Rep's Initials
_____ Trade Contractor's Initials

officials. Such requirements shall take precedence over any item mentioned in these specifications and standards if they are in conflict. Installations should pass the initial compliance inspection at each phase.

Wiring shall be capable of carrying its designed load under normal residential use for the first two (2) years of the Warranty period.

Materials: The electrical system shall be installed using new materials of the grade and quality specified or required to meet the above standards.

Installation: All work is to be done by trained, experienced individuals. Workmanship shall be sufficient so that a minimum of repairs are required after installation. **Under no circumstances shall any roof or floor truss be cut, notched, or damaged.** Should any truss be accidentally damaged the Site Superintendent should be notified immediately. No repairs may be made without the approval of the truss manufacturer's engineer.

Warranty: Certain items are warrantable for one (1) year and some for two (2) years. Refer to The Company's printed Limited Warranty booklet for a list of these items. The Trade Contractor shall furnish to the homeowner, via a tag on the breaker box and a sticker on the inside of the top left door of the cabinet located to the left of the kitchen sink, a regular office hour phone number and an emergency phone number. The Trade Contractors is required to furnish emergency service.

The Trade Contractor shall have seven (7) days in which to correct any Warranty problem. If the problem is not corrected within seven (7) days then The Company shall correct the problem and will backcharge the Trade Contractor at the rate of $25.00 per hour, with a minimum charge of $100.00 plus the cost of any materials.

Inspection Reports: The Trade Contractor and the Site Superintendent shall walk the job together and complete each section of the inspection report(s). The Trade Contractor must correct any deficiency found during the inspection and the job must be 100-percent complete before payment will be made. The Trade Contractor and the Site Superintendent must sign-off on all sections of the inspection report(s) attesting that the job is 100-percent complete and is correct per the job requirements found in this Scope of Work.

Detailed Job Requirements:

1. A new set of plans is required for each house. Plans are subject to changes and modifications. It is the responsibility of the Trade Contractor to have new plans before beginning work. Plans should be picked up at the job trailer from the Site Superintendent. The Trade Contractor at no cost to The Company will correct any errors that occur from using an incorrect set of plans.

2. A copy of the purchase order will be mailed from the office. If one is not received a copy can be picked up from the Site Superintendent.

3. The Trade Contractor and the Site Superintendent must walk the job together and complete the pre-work section of the inspection report(s) before work may begin. The pre-work section of the inspection report(s) must be signed-off on by both parties.

Company Rep's Initials _____
Trade Contractor's Initials _____

4. The Trade Contractor must be licensed in the county in which he or she will be working before work may begin.

5. The Trade Contractor is responsible for securing all necessary permits.

6. The Trade Contractor is responsible for setting a temporary power pole and securing temporary power.

7. All electrical work shall meet or exceed the requirements of the latest edition of the National Electrical Code (NEC) and/or meet or exceed any local codes.

8. A breaker box of a size that will adequately handle all electrical requirements of the house shall be installed. Circuit breakers shall be trip-free and capable of being closed and opened by manual operation.

9. All circuit breakers shall be plainly and legibly labeled.

10. Receptacle outlets shall be installed per NEC code and shall be, insofar as practicable, spaced equal distances.

11. Additional outlets (over and above the NEC code requirements) will be noted on the purchase order and/or change order.

12. All receptacle outlets shall be installed so that they do not protrude past the drywall. All receptacle covers shall fit level, square, plumb, and snugly against the wall.

13. Receptacle outlets in the kitchen, garage, and baths shall be a GFI type. The number will be determined by the NEC code.

14. Wiring and bracing for ceiling fans will be as indicated on the plans. The Trade Contractor shall install bracing for ceiling fans in all bedrooms and bonus rooms, even if a ceiling fan is not installed.

15. If the plans call for a bedroom with a ceiling fan a switched receptacle also must be provided.

16. The Trade Contractor shall furnish and install outside floodlights, bathroom fans, doorbells, and smoke detectors per plans.

17. All smoke detectors shall be electric and also equipped with backup battery power.

18. The Trade Contractor is responsible for the installation of all light fixtures per plan, including the installation of light bulbs in all fixtures. The Company shall furnish light fixtures and bulbs.

19. In any bath with a two-basin sink one GFI wall receptacle outlet shall be installed adjacent to each basin location.

20. All houses shall have two outside receptacles, one in the front and one in the back of the house.

21. The Trade Contractor is responsible for cutting neat, clean holes for wiring in either vinyl or concrete siding. All holes must be sealed and all excess sealing material removed from the siding.

_____ Company Rep's Initials
_____ Trade Contractor's Initials

22. All laundry areas will include a 220-volt receptacle outlet.

23. Wiring shall be provided for all heating and air conditioning (HVAC) units. When a unit is located in the attic the Trade Contractor shall install a permanent electric outlet and lighting fixture near the equipment, which shall be controlled by a switch located at the passageway opening. All heating units shall have a permanent electric outlet and lighting fixture located in such a manner as to illuminate the unit for maintenance, filter changes, etc.

24. Thermostats shall be wired per plan.

25. When a light is being installed over the kitchen sink the Trade Contractor shall bring the wiring through the wall as close to the top of the cabinet as possible and run the wire along the top of the cabinet to the fixture location. The wire for the light should not be visible coming through the wall when standing and looking at the wall above the cabinet.

26. The correct and necessary wiring shall be provided for all appliances per plan.

27. Outside disconnects shall be wired per county code as required.

28. Trade Contractor is responsible for passing all required inspections.

29. Any reinspection fees assessed The Company will be charged to the Trade Contractor.

30. All construction debris must be removed to the dumpster or to an area designated by the Site Superintendent. The job will not be considered complete and payment will not be issued until all trash and debris have been removed from the house and/or site.

31. The Trade Contractor and Site Superintendent must walk the job together and perform a final inspection of the job. The final section of the inspection report(s) must be completed and signed-off on by both parties. The inspection report(s) must be attached to the office's copy of the purchase order and the Trade Contractor's invoice or payment will not be issued.

32. Any items found during any inspection that need correction shall be corrected before payment will be made.

I _____ agent for _____

_____ have read and fully understand the above **Scope of Work** and
I hereby agree to perform all work in accordance with the above.

Date: _____   _____
                            Signed: Trade Contractor (or agent)

Date: _____   _____
                            For The Company

INSPECTION REPORT

ELECTRIC (ROUGH)

Subdivision/Lot # _____

Pre-Work Inspection: (to be completed prior to beginning work)

____   Plans have been picked up and signed for.

____   Purchase order has been received.

____   Tubs, sinks, etc., are covered and protected, and are free of damage.

____   Temporary handrails, if required, are installed.

____   Safety bracing is in place as needed.

____   If house requires two HVAC units, the attic access is ready for wiring.

____   House is roofed and shingled.

____   House is broom-swept, clean, and free of debris.

Date: _____   Site Superintendent _____

Trade Contractor _____

Final Inspection: (to be completed before Trade Contractor leaves jobsite)

____   Temporary power pole is set and temporary power is on.

____   Wiring is complete for light switches and receptacles per plan.

____   All receptacle boxes are set square, plumb, and 3/8 inch past stud.

____   Bracing for ceiling fans, as required, is installed.

____   Wiring for thermostats is complete.

____   Wiring for smoke detectors, washer, dryer, flood lights, etc., is complete.

____   Wiring for HVAC units is complete.

____   Wiring for outside receptacles is complete.

____   Outside disconnects are installed.

____   Breaker box is installed.

____   Breakers are clearly and legibly labeled.

____   Emergency stickers are in place on breaker box.

____   Inspection has been passed.

____   Site and house is clean and free of debris.

Date: _____   Site Superintendent _____

Trade Contractor _____

INSPECTION REPORT

ELECTRIC (FINAL)

Subdivision/Lot # _____

Pre-Work Inspection: (to be completed prior to beginning work)

____ Purchase order has been received.

____ House is clean and free of debris.

____ Tubs, sinks, etc., are covered and protected, and free of damage.

____ Drywall is complete with no damage.

____ Painting is complete with no damage.

____ All receptacle boxes are clean and free of drywall mud and other debris.

____ House is at stage for final electrical to be completed.

____ House is broom-swept, clean, and free of debris.

Date: _____   Site Superintendent _____

                              Trade Contractor _____

Final Inspection: (to be completed before Trade Contractor leaves jobsite)

____ All faceplates are installed square, plumb, and snug against walls.

____ All receptacles are tested and all work as required.

____ All appliances are hooked up and tested.

____ Garbage disposal is tested.

____ All light fixtures are installed and tested.

____ Smoke detectors are hooked up and tested.

____ Outside disconnect has been tested.

____ HVAC units are hooked up and tested.

____ Bath fans and ceiling fans are hooked up and tested.

____ Door bell is hooked up and tested.

____ Wire for light over kitchen sink is not visible above lip of top cabinet.

____ Inspection has been passed.

____ House is clean and free of debris.

____ Site is clean and free of debris.

Date: _____   Site Superintendent _____

                              Trade Contractor _____

SCOPE OF WORK
Standards and Description of Work Performance

FIREPLACES

The Company's **Terms and Conditions** are by reference a part of all **Scope of Work** requirements.

Construction Requirements:

Generally speaking the work of Trade Contractors and their employees is expected to be performed in a good and workmanlike manner. Workmanlike quality is defined as workmanship that meets or betters those criteria indicated in applicable building codes, using materials and installation methods identified in the construction plans and this Scope of Work.

Code Requirements:

All jobs shall conform to those standards stipulated in the building code, mechanical code, plumbing code, and electrical code applicable in the local jurisdiction. All construction on The Company's jobsites shall meet or exceed NAHB Performance and Building Standards.

General Comments:

The Company considers our Trade Contractors to be experts at producing a high-quality job. But everyone on our construction team—staff, Trade Contractors, and suppliers—must recognize the importance of providing quality in both the product and service areas while on our jobsites and in the homes of our purchasers.

Since we work as a team, poor quality or service, from any of us, reflects unfavorably on all of us. An exceptional level of product quality and highly effective service can help us all to increase our business and grow.

The Company's definition of quality construction also requires that every job be completed correctly the first time. When this does not occur it costs both of us additional money, imposes on the purchaser, and hurts our reputations as quality builders. That is why, in situations where construction was not completed in a quality manner, prompt corrective action is required to remedy specific deficiencies.

In the following information the term Site Superintendent shall refer to any Company representative with authority to perform the specified task. The term Trade Contractor shall mean the Trade Contractor's organization or any representative that is assigned the authority to perform the specified task.

General: The fireplace shall be the size and type designated on the purchase order and house plans. It shall be a prefabricated unit and shall be installed by the Trade Contractor per the manufacturer's installation instructions.

_____ Company Rep's Initials
_____ Trade Contractor's Initials

Installation: All installation shall be done by trained, experienced individuals. All fireplaces shall be installed level, plumb, and square, and must meet or exceed all fire and county, state, or local codes. No fireplace box may be out of line more than 1/4 inch. The fireplace should be installed in such a manner that when the face of the fireplace is installed the face shall fit snug against all walls. During installation care must be taken by the Trade Contractor not to damage the work of other trade contractors.

Warranty: The Company believes that all work done and all materials installed, in connection with one of our homes should be of high quality and that all Trade Contractors should stand behind the quality of their work and materials. Therefore we require all Trade Contractors to warrant the quality of their work for a period of one (1) year from date of closing of the house. Please refer to our printed Limited Warranty booklet for specific items that are covered under Warranty as they apply to the fireplace(s).

The Trade Contractor shall have seven (7) days in which to correct any Warranty problem. If the problem is not corrected within seven (7) days then The Company shall correct the problem and will backcharge the Trade Contractor at the rate of $25.00 per hour, with a minimum charge of $100.00 plus the cost of any materials.

Inspection Reports: The Trade Contractor and the Site Superintendent shall walk the job together and complete each section of the inspection report(s). The Trade Contractor must correct any deficiency found during the inspection and the job must be 100-percent complete before payment will be made. The Trade Contractor and the Site Superintendent must sign-off on all sections of the inspection report(s) attesting that the job is 100-percent complete and is correct per the job requirements found in this Scope of Work.

Detailed Job Requirements:

1. The Trade Contractor and the Site Superintendent must walk the job together and complete the pre-work section of the inspection report(s) before work may begin. The pre-work section of the inspection report(s) must be signed-off on by both parties.

2. The purchase order will be mailed from the office to the Trade Contractor. If it is not received, a copy may be picked up from the Site Superintendent.

3. All fireplaces will be the type and size specified on the purchase order.

4. Fireplaces shall be installed per plans.

5. The Trade Contractor is responsible for complying with all fire, state, local, building, and other applicable codes.

6. The fireplace must be installed level, plumb, and square. No fireplace may be out of line more than 1/4 inch.

7. It is the responsibility of the Trade Contractor to install deadwood as necessary.

Company Rep's Initials _____
Trade Contractor's Initials _____

8. The framing Trade Contractor should have installed all firestops. In the event the Trade Contractor finds that firestops are missing the Site Superintendent should be notified immediately.

9. The Trade Contractor shall not damage another trade contractor's work. If any wood needs to be cut the Trade Contractor should not damage the subfloor or any framing member.

10. All chimneys shall be capped or a shroud installed per plans.

11. All fireplaces shall be equipped with gas starters.

12. If the Trade Contractor furnishes and installs the fireplace fronts, hearths, and mantels, they shall be installed per the manufacturer's installation instructions.

13. All fireplace faces, hearths, and mantles shall be installed square, level, and plumb. The face and/or mantel shall fit smoothly and snugly against the walls. There shall be no more than a 1/8-inch gap to be caulked on any side.

14. The fireplace key, escutcheons, and information manuals shall be delivered to The Company's office. These will be included in the New Homeowner's Manual given to the homeowner prior to closing.

15. Any trash or debris shall be removed to the dumpster and the site and house must be clean before the job will be considered complete.

16. The Trade Contractor and Site Superintendent must walk the job together and perform a final inspection of the job. The final section of the inspection report(s) must be completed and signed-off on by both parties. The inspection report(s) must be attached to the office's copy of the purchase order and the Trade Contractor's invoice or payment will not be issued.

17. Any items found during the final inspection that need correction shall be corrected before payment will be made.

I _____ agent for _____

_____ have read and fully understand the above **Scope of Work** and I hereby agree to perform all work in accordance with the above.

Date: _____   _____
                          Signed: Trade Contractor (or agent)

Date: _____   _____
                          For The Company

INSPECTION REPORT

FIREPLACES (ROUGH)

Subdivision/Lot # _____

Pre-Work Inspection: (to be completed prior to beginning work)

____ Plans have been reviewed.

____ Purchase order has been received.

____ Room(s) in which fireplace is to be installed are clean and free of debris.

____ Framing is complete.

____ There are no cuts in the subfloor.

Date: _____   Site Superintendent _____

Trade Contractor _____

Rough Inspection: (to be completed before Trade Contractor leaves jobsite)

____ Fireplace is installed level, plumb, and square.

____ Fireplace does not extend more than 3/8 inch on any side.

____ There are no cuts in the subfloor.

____ Trash is removed from the house to dumpster or designated area.

Date: _____   Site Superintendent _____

Trade Contractor _____

INSPECTION REPORT

FIREPLACES (FINAL)

Subdivision/Lot # _____

Pre-Work Inspection: (to be completed prior to beginning work)

____   Purchase order has been received.

____   Room(s) in which fireplace(s) is to be installed are clean and free of debris.

____   Fireplace is level.

____   Drywall is complete.

Date: _____   Site Superintendent _____

                                Trade Contractor _____

Final Inspection: (to be completed before Trade Contractor leaves jobsite)

____   Fireplace is still level, plumb, and square.

____   Screen is in place.

____   Flue opens and closes correctly.

____   Fresh air vent works.

____   Face installed correctly—level, plumb, and square.

____   Gap around the face is no more than 1/8 inch.

____   Mantle is level, square, and snug with no more than 1/8 inch caulked.

____   Hearth is installed level, square, and plumb/flat floor—flush with floor.

____   Gas starter checked and working.

____   Gas starter works and is checked.

____   Cap or shroud is installed on chimney.

____   House is clean and free of debris.

Date: _____   Site Superintendent _____

                                Trade Contractor _____

<u>SCOPE OF WORK</u>
Standards and Description of Work Performance

FLOOR COVERINGS

The Company's **Terms and Conditions** are by reference a part of all **Scope of Work** requirements.

Construction Requirements:

Generally speaking the work of Trade Contractors and their employees is expected to be performed in a good and workmanlike manner. Workmanlike quality is defined as workmanship that meets or betters those criteria indicated in applicable building codes, using materials and installation methods identified in the construction plans and this Scope of Work.

Code Requirements:

All jobs shall conform to those standards stipulated in the building code, mechanical code, plumbing code, and electrical code applicable in the local jurisdiction. All construction on The Company's jobsites shall meet or exceed NAHB Performance and Building Standards.

General Comments:

The Company considers our Trade Contractors to be experts at producing a high-quality job. But everyone on our construction team—staff, Trade Contractors, and suppliers—must recognize the importance of providing quality in both the product and service areas while on our jobsites and in the homes of our purchasers.

Since we work as a team, poor quality or service, from any of us, reflects unfavorably on all of us. An exceptional level of product quality and highly effective service can help us all to increase our business and grow.

The Company's definition of quality construction also requires that every job be completed correctly the first time. When this does not occur it costs both of us additional money, imposes on the purchaser, and hurts our reputations as quality builders. That is why, in situations where construction was not completed in a quality manner, prompt corrective action is required to remedy specific deficiencies.

In the following information the term Site Superintendent shall refer to any The Company representative with authority to perform the specified task. The term Trade Contractor shall mean the Trade Contractor's organization or any representative that is assigned the authority to perform the specified task.

General: All carpet, pad, vinyl, parquet, or hardwood flooring shall be of a type, brand, and grade specified and approved by The Company. These standards may not be changed without the written approval of The Company's purchasing agent and the change must be approved in writing. If the

Company Rep's Initials _____
Trade Contractor's Initials _____

type, brand, or grade is changed the Trade Contractor must furnish to The Company Warranty and other information on the item for inclusion in The Company's New Homeowner's Manual.

Materials: The type, grade, color, etc., of all carpet, vinyl, parquet, and hardwood flooring shall be furnished to the Trade Contractor two (2) weeks prior to the installation of these items. There shall be no substitution of type, grade, quality, or manufacturer of any floor covering.

Installation: All installations shall be done by trained, experienced personnel. No carpet shall buckle, wrinkle, or otherwise not present a smooth, professional appearance. Seam lines are unavoidable, but there shall be no visible gaps in any floor covering. Carpets shall be flush with the inside of the tack strip. No visible gap around any wall or baseboard is acceptable.

All floor coverings shall be installed per the manufacturer's instructions, using materials, fasteners, adhesives, etc., as recommended by the manufacturer.

In this nonperfect world not every wall is perfect. Patterned vinyl, parquet, or hardwood flooring should be positioned to minimize any wall deflection. Nothing turns a homeowner off more than to walk into a bath or kitchen and immediately notice that a wall is slightly off. The installer can help minimize this by adjusting the pattern slightly in the most visible areas.

Turnkey Job: The Trade Contractors shall furnish a turnkey job. This job shall include the preparation of the floor, the installation of underlayment, the installation of the floor covering itself, and full Warranty on all materials and installation.

Warranty: The Company believes that all work done and all materials installed, in connection with one of our homes should be of high quality and that all Trade Contractors should stand behind the quality of their work and materials. Therefore we require all Trade Contractors to warrant the quality of their work for a period of one (1) year from date of closing of the house. Please refer to our printed Limited Warranty booklet for specific items that are covered under Warranty as they apply to the floor-covering Trade Contractor.

The Trade Contractor shall have seven (7) days in which to correct any Warranty problem. If the problem is not corrected within seven (7) days then The Company shall correct the problem and will backcharge the Trade Contractor at the rate of $25.00 per hour, with a minimum charge of $100.00 plus the cost of any materials.

Inspection Reports: The Trade Contractor and the Site Superintendent shall walk the job together and complete each section of the inspection report(s). The Trade Contractor must correct any deficiency found during the inspection and the job must be 100-percent complete before payment will be made. The Trade Contractor and the Site Superintendent must sign-off on all sections of the inspection report(s) attesting that the job is 100-percent complete and is correct per the job requirements found in this Scope of Work.

Detailed Job Requirements:

1. A new set of plans is required for each house. Plans are subject to changes and modifications. It is the responsibility of the Trade Contractor to have the new plans before beginning work. Plans

_____ Company Rep's Initials
_____ Trade Contractor's Initials

should be picked up at the job trailer from the Site Superintendent. The Trade Contractor at no cost to The Company will correct any errors that occur from using an incorrect set of plans.

2. Purchase orders will be mailed to the Trade Contractor.

3. Color selections shall be mailed to the Trade Contractor a minimum of two (2) weeks prior to the beginning of the job.

4. The Trade Contractor and the Site Superintendent must walk the job together and complete the pre-work section of the inspection report(s) before work may begin. The pre-work section of the inspection report(s) must be signed-off on by both parties.

5. All floors will be fully prepped and clean before any floor covering is installed.

6. If any floor is too damp, is not level, has large cracks, or is otherwise unsuitable for the installation of the floor coverings the Site Superintendent should be notified immediately.

7. The Trade Contractor shall apply underlayment, floor-leveling compound, adhesive, etc., as required to achieve a professional, high-quality finished product.

8. No point on the surface of a wood floor shall be more than 1/2 inch higher or lower than any other point on the surface within 20 feet or proportional multiples of the preceding dimensions.

9. All parquet and hardwood flooring shall be installed without bows or bubbles, and shall be laid square, plumb, and per the manufacturer's installation instructions. Gaps between strips of hardwood floorboards shall not exceed 1/8 inch in width at time of installation. Deflections and gaps in parquet and hardwood flooring shall not have more than a 1/4-inch ridge or depression within any 32-inch measurement.

10. All blocks of parquet flooring shall be as close to the same color as possible. Blocks of an obvious color difference will not be accepted. Final inspection of all parquet or hardwood flooring shall be done in sunlight and also in normal house lighting. Parquet and hardwood flooring cannot display any defects that can be readily seen at a distance of 6 feet under these conditions.

11. The contractor shall install the grade of hardwood as specified by the purchase order. All wood should be consistent with grading stamp as specified.

12. Slivers or splinters that occur during the installation of the flooring are unacceptable.

13. Crowning in strip flooring shall not exceed 1/16 inch in depth in a 3-inch maximum span when measured perpendicular to the long axis of the board.

14. Parquet and hardwood flooring must be firmly attached to the floor. It shall not bubble or come loose within the Warranty period.

15. All parquet and hardwood flooring shall be cut to fit 1/8 inch of baseboard.

16. Parquet and hardwood shall be covered to protect it from damage during the remaining construction of the house.

Company Rep's Initials _____
Trade Contractor's Initials _____

17. Vinyl shall be **run** under all door casing. Cutting around door facings is not acceptable. No cuts shall be made that allows the underlayment or slab to be visible under any casing.

18. Vinyl shall not be cut around commodes. Vinyl shall be pushed under the commode or the commode shall be lifted and vinyl placed so that no cut areas are visible when the commode is reset.

19. All vinyl shall be cut to fit flush to baseboards.

20. Vinyl shall be cut large enough at all ducts so that when the grill is installed no gaps or cuts are visible.

21. Patterned vinyl shall be laid in accordance with wall deflections to reduce visual deflection of the wall against the pattern.

22. Patterns at seams between adjoining pieces shall be aligned to within 1/8 inch.

23. Vinyl shall be covered to protect it from damage during the remaining construction of the house.

24. Vinyl shall be firmly attached to the underlayment or slab. It shall not bubble, show nail pops, or come loose from the adhesive during the Warranty period.

25. All vinyl adhesive shall be thoroughly cleaned from vinyl before the job will be considered complete.

26. All vinyl shall be inspected in both sunlight and normal lighting. Any defects visible from a distance of 6 feet under these conditions are unacceptable.

27. Carpet pad shall be installed per the manufacturer's installation instructions and shall be of the grade and type specified on the purchase order.

28. Carpet shall be installed per the manufacturer's installation instructions with the fasteners approved by the manufacturer.

29. Carpet shall be installed flush with the inside edge of the tack strip. Any short-cut carpet in wall or foyer areas will be replaced. Patching is not acceptable.

30. Carpet shall be firmly attached to the tack strip and shall not come loose within the Warranty period. Tack strips shall be firmly attached per the manufacturer's installation instructions.

31. Carpet shall be butted firmly against parquet or hardwood flooring.

32. Transition strips shall be used at the junction of carpet and vinyl. The strips shall be firmly anchored. Lippage greater than 1/16 inch is considered excessive.

33. Seams shall be as inconspicuous as possible. Gaps are unacceptable.

34. All carpet shall be inspected in both sunlight and normal lighting. Any defects visible from a distance of 6 feet under these conditions are unacceptable.

_____ Company Rep's Initials
_____ Trade Contractor's Initials

35. The Trade Contractor shall furnish heaters if required. If heat is required overnight the Trade Contractor is not required to leave his or her own heaters in the house, but must notify the Site Superintendent of the need for heat overnight.

36. The Trade Contractor shall be responsible for any shrinkage or shortage of the pad during the first year of Warranty coverage.

37. The Trade Contractor and Site Superintendent must walk the job together and perform a final inspection of the job. The final section of the inspection report(s) must be completed and signed-off on by both parties. The inspection report must be attached to the office's copy of the purchase order and the Trade Contractor's invoice or payment will not be issued.

38. Any items found during the final inspection that need correction shall be corrected before payment will be made.

I _____ agent for _____

_____ have read and fully understand the above **Scope of Work** and I hereby agree to perform all work in accordance with the above.

Date: _____

_____
Signed: Trade Contractor (or agent)

Date: _____

_____
For The Company

INSPECTION REPORT

FLOOR COVERINGS

Subdivision/Lot # _____

Initial Inspection (prior to installing parquet, hardwood, or vinyl flooring): (to be completed prior to beginning work)

____ Plans have been verified.

____ Purchase order has been received.

____ Color selection has been verified.

____ Jobsite is clean and free of debris.

____ House is clean and broom-swept.

____ Driveway is clean and barricaded.

____ Temporary handrails, bracing, etc., are in place.

____ Floor is level with no more than 1/4 inch in 32-inches deviation.

____ Floor is clean.

____ Subflooring cracks are filled and sanded level.

____ No bumps, overlaps, etc., are in subfloor or slab.

____ Subflooring or slab is dry.

Date: _____   Site Superintendent _____

Trade Contractor _____

Final Inspection (after installing parquet, hardwood, or vinyl flooring): (to be completed before Trade Contractor leaves jobsite)

____ Floors are level after installation of floor covering with no deviation more than 1/4 inch in 32 inches.

____ Floor coverings are cut within 1/8 inch of wall.

____ Floor coverings are firmly attached with no bubbles, nail pops, lumps, bumps, etc.

____ Pattern of vinyl is laid correctly and in a manner that helps diminish any wall deflections.

____ Floor coverings are correct at door casings. Subflooring or slab is not visible.

____ Floor coverings are correct at toilets.

____ Vinyl is flush with cabinets and tubs.

____ Floor coverings are correct at all ducts.

____ Parquet and/or hardwood flooring does not have gaps, splinters, or loose areas.

____ Parquet, hardwood, and vinyl flooring has been inspected under normal interior lighting and also in sunlight from windows. No defects were found.

____ Floor coverings are protected from damage.

____ House is clean with all working areas broom-swept.

____ Debris and trash have been removed to designated trash site.

____ Left-over vinyl and parquet/hardwood flooring has been stacked in the garage, store room, or laundry area as instructed by the Site Superintendent.

Date: _____   Site Superintendent _____

Trade Contractor _____

INSPECTION REPORT

FLOOR COVERING (CARPET)

Subdivision/Lot # _____

Initial Inspection (prior to installing carpet): (to be completed prior to beginning work)

____   House is clean and free of debris.

____   Drive is barricaded.

____   House has been rough cleaned.

____   Windows have been cleaned.

____   Floors are clean and ready to be prepped.

____   Hardwood/parquet and vinyl flooring is clean and undamaged.

____   Drywall is complete.

____   Painting is complete.

____   Cabinets, toilets, etc., are all installed.

____   Light fixtures are installed.

____   Switch and receptacle covers are in place.

____   All hardware is installed.

____   Garage is clean, broom-swept, and ready to be used to cut carpet.

____   HVAC is working.

Date: _____   Site Superintendent _____

Trade Contractor _____

Final Inspection (after installing carpet): (to be completed before Trade Contractor leaves jobsite)

____   Floor is level with no underlying lumps, bumps, hilly areas, depressions, etc.

____   Carpet is cut to within 1/8 inch of wall.

____   Transition strips are installed between vinyl and carpet and firmly seated.

____   Pad and carpet are firmly attached to tack strip.

____   Carpet is butted firmly against parquet and hardwood flooring.

____   No gaps are visible.

____   Seams are very inconspicuous.

____   Carpet is well stretched so that it is tight and has no wrinkles.

____ No obvious color variations are visible.

____ Carpet has been inspected under normal interior lighting and also in sunlight from windows.

____ Left-over carpet is in garage, store room, or laundry area.

____ House is clean and free of debris.

____ Jobsite is clean.

Date: _____   Site Superintendent _____

Trade Contractor _____

<u>SCOPE OF WORK</u>
Standards and Description of Work Performance

FOOTINGS

The Company's **Terms and Conditions** are by reference a part of all **Scope of Work** requirements.

Construction Requirements:

Generally speaking the work of Trade Contractors and their employees is expected to be performed in a good and workmanlike manner. Workmanlike quality is defined as workmanship that meets or betters those criteria indicated in applicable building codes, using materials and installation methods identified in the construction plans and this Scope of Work.

Code Requirements:

All jobs shall conform to those standards stipulated in the building code, mechanical code, plumbing code, and electrical code applicable in the local jurisdiction. All construction on The Company's jobsites shall meet or exceed NAHB Performance and Building Standards.

General Comments:

The Company considers our Trade Contractors to be experts at producing a high-quality job. But everyone on our construction team—staff, Trade Contractors, and suppliers—must recognize the importance of providing quality in both the product and service areas while on our jobsites and in the homes of our purchasers.

Since we work as a team, poor quality or service, from any of us, reflects unfavorably on all of us. An exceptional level of product quality and highly effective service can help us all to increase our business and grow.

The Company's definition of quality construction also requires that every job be completed correctly the first time. When this does not occur it costs both of us additional money, imposes on the purchaser, and hurts our reputations as quality builders. That is why, in situations where construction was not completed in a quality manner, prompt corrective action is required to remedy specific deficiencies.

In the following information the term Site Superintendent shall refer to any The Company representative with authority to perform the specified task. The term Trade Contractor shall mean the Trade Contractor's organization or any representative that is assigned the authority to perform the specified task.

General—Concrete: Concrete is subject to various phenomena including applied chemicals and natural elements that can deteriorate the surface. In some cases however, surface defects such as flaking, scaling, or spallling may be caused by improper finishing. Excessive powdering or chalking may occur due to improper troweling, excess water, or when uncured concrete is allowed to freeze.

_____ Company Rep's Initials
_____ Trade Contractor's Initials

It is important that proper precautions and correct techniques be utilized in the handling and finishing of concrete in the field to ensure a quality job.

Foundation and structural concrete shall maintain its integrity. Poured concrete should not crack in excess of 3/16 inch in vertical or horizontal displacement. Concrete shall not pit, scale, or spall to the extent that the aggregate is exposed under normal weathering and use.

Installation: Concrete should be placed at an appropriate rate so that it can be spread, straightened, and darbied properly. Techniques for handling and placing concrete should ensure that it remains uniform within each batch and from batch to batch. Concrete should not be allowed to run or be worked over long distances, and should not be allowed to drop more than 4 feet.

Finish: The top surface of footings shall be level, all corners shall be square, and all footings shall be straight. The bottom surface of footings may not have a slope exceeding 1 inch in 10 inches. Footings shall be stepped where necessary to change the elevation of the top surface of the footings or where the slope of the bottom surface of the footing will exceed 1 inch in 10 inches.

Footings shall extend not less than 12 inches below the finished natural grade or engineered fill, and in no case less than the frostline depth. Footing sizes are based on soil with an allowable soil pressure of 2,000 pounds per square foot. Footings on soil with a lower allowable soil pressure shall be designed in accordance with accepted engineering practice. Footing projections shall not exceed the footing thickness.

If piers are required they shall be a minimum of 2 feet in solid, undistributed soil, and a minimum of 6 feet apart unless an engineer's report requires different placements.

The following shall be the size of all footings unless designated differently by an engineered footing report:

- Single-story-home footings shall be 16 inches wide by 8 inches deep.
- Two-story-home or single-story-on-basement footings shall be 16 inches wide by 8 inches deep.
- Three-story-home or two-story-on-basement footings shall be 18 inches wide by 8 inches deep.

Warranty: The Company believes that all work done in connection with one of our homes should be of high quality and that all Trade Contractors should stand behind the quality of their work. Therefore we require all Trade Contractors to warrant the quality of their work for a period of one (1) year from date of closing of the house. Please refer to our printed Limited Warranty booklet for any specific items that are covered under Warranty as they apply to the footing Trade Contractor.

The Trade Contractor shall have seven (7) days in which to correct any Warranty problem. If the problem is not corrected within seven (7) days then The Company shall correct the problem and will backcharge the Trade Contractor at the rate of $25.00 per hour, with a minimum charge of $100.00 plus the cost of any materials.

Inspection Reports: The Trade Contractor and the Site Superintendent shall walk the job together and complete each section of the inspection report(s). The Trade Contractor must correct any deficiency found during the inspection and the job must be 100-percent complete before payment will be made. The Trade Contractor and the Site Superintendent must sign-off on all sections of the

Company Rep's Initials _____
Trade Contractor's Initials _____

inspection report(s) attesting that the job is 100-percent complete and is correct per the job requirements found in this Scope of Work.

Detailed Job Requirements:

1. A new set of plans is required for each house. Plans are subject to changes and modifications. It is the responsibility of the Trade Contractor to have the new plans before beginning work. Plans should be picked up at the job trailer from the Site Superintendent. The Trade Contractor at no cost to The Company will correct any errors that occur from using an incorrect set of plans.

2. Purchase orders should be picked up from the Site Superintendent.

3. Prior to beginning to dig footings the Trade Contractor and the Site Superintendent shall use the foundation plans to verify all the dimensions of the house, that all stakes are in place, that all offsets are staked, and that all lines are straight and true.

4. The Trade Contractor shall verify with the Site Superintendent that the staked area of the house will fit on the lot within the building lines and any easements.

5. The Trade Contractor and the Site Superintendent must walk the job together and complete the pre-work section of the inspection report(s) before work may begin. The pre-work section of the inspection report(s) must be signed-off on by both parties.

6. All footings for stoops and porches shall be dug and poured as a part of the foundation for the house.

7. The Trade Contractor shall dig all footings. Footings may be dug by machine or by hand. They should not be over-dug, and must be of width and depth to accommodate the amount of concrete listed below.

8. Footings for block foundation walls shall be dug, shaped, and concrete poured 16 inches wide by 8 inches deep for single-story homes; 16 inches wide by 8 inches deep for a two-story homes or a single-story house on a basement; or 18 inches wide by 8 inches deep for a three-story house or two-story house on a basement, unless an engineer's report requires a different size.

9. Two #4 rebar shall be installed continuously in all footings half way through the depth of the footing. Bending and cutting of reinforcing steel is to be done by the Trade Contractor. Additional rebar may be required for engineered footings.

10. All concrete shall be 3,000 psi.

11. The Trade Contractor is responsible for cleaning up all debris. Any and all concrete runoff shall be removed from the site and placed in the driveway cut.

12. The Trade Contractor and Site Superintendent must walk the job together and perform a final inspection of the job. The final section of the inspection report(s) must be completed and signed-off on by both parties. The inspection report(s) must be attached to the office's copy of the purchase order and the Trade Contractor's invoice or payment will not be issued.

_____ Company Rep's Initials
_____ Trade Contractor's Initials

13. Any items found during the final inspection that need correction shall be corrected before payment will be made.

I _____ agent for _____

_____ have read and fully understand the above **Scope of Work** and
I hereby agree to perform all work in accordance with the above.

Date: _____

_____
Signed: Trade Contractor (or agent)

Date: _____

_____
For The Company

INSPECTION REPORT

FOOTINGS

Subdivision/Lot # _____

Pre-Work Inspection: (to be completed prior to beginning work)

____ Plans have been picked up and signed for.

____ Purchase order has been picked up from Site Superintendent.

____ Work area is clean and free of debris.

____ Site is cleared and pad cut.

____ House is spotted, stakes set, and dimensions match plans.

____ House fits on lot within building and easement lines.

____ Straightness (alignment) of lines has been verified.

____ All setbacks, bay windows, fireplaces, porches, stoops, etc., are clearly marked and dimensions checked.

____ Area is cleared and ready for cement truck access.

____ Driveway is cut and filled with stone.

Date: _____   Site Superintendent _____

Trade Contractor _____

Final Inspection: (to be completed before Trade Contractor leaves jobsite)

____ Depth and width of concrete is correct.

____ Corners are square.

____ Footings are level and plumb.

____ Footings are straight with no more than a 4-inch deviation in 20 feet.

____ All setout or setback dimensions have been verified.

____ Overall dimensions have been verified against plans.

____ No aggregate is exposed in any area.

____ Site is clean and free of excess concrete.

____ Foundation measurements have been verified.

Date: _____   Site Superintendent _____

Trade Contractor _____

<u>SCOPE OF WORK</u>
Standards and Description of Work Performance

FRAMING LABOR

The Company's **Terms and Conditions** are by reference a part of all **Scope of Work** requirements.

Construction Requirements:

Generally speaking the work of Trade Contractors and their employees is expected to be performed in a good and workmanlike manner. Workmanlike quality is defined as workmanship that meets or betters those criteria indicated in applicable building codes, using materials and installation methods identified in the construction plans and this Scope of Work.

Code Requirements:

All jobs shall conform to those standards stipulated in the building code, mechanical code, plumbing code, and electrical code applicable in the local jurisdiction. All construction on The Company's jobsites shall meet or exceed NAHB Performance and Building Standards.

General Comments:

The Company considers our Trade Contractors to be experts at producing a high quality job. But everyone on our construction team—staff, Trade Contractors, and suppliers—must recognize the importance of providing quality in both the product and service areas while on our jobsites and in the homes of our purchasers.

Since we work as a team, poor quality or service, from any of us, reflects unfavorably on all of us. An exceptional level of product quality and highly effective service can help us all to increase our business and grow.

The Company's definition of quality construction also requires that every job be completed correctly the first time. When this does not occur it costs both of us additional money, imposes on the purchaser, and hurts our reputations as quality builders. That is why, in situations where construction was not completed in a quality manner, prompt corrective action is required to remedy specific deficiencies.

In the following information the term Site Superintendent shall refer to any The Company representative with authority to perform the specified task. The term Trade Contractor shall mean the Trade Contractor's organization or any representative that is assigned the authority to perform the specified task.

General: Floor system variance should not exceed 1/2 inch out of level in 20 feet, with no ridges or depressions in excess of 1/4 inch within any 32-inch measurement. Floor trusses shall be installed per the manufacturer's layout.

Company Rep's Initials _____
Trade Contractor's Initials _____

Framing and sheathing members shall be fastened according to code and manufacturer's instructions.

Studs, plates, cripples, and headers shall be cut squarely and within 1/8 inch of required length. Wall layout measurement variance shall not exceed 1/4 inch in 20 feet. Bearing walls should not exceed 1/2 inch out of level in 20 feet, with not more than a 1/4-inch ridge or depression within any 32-inch measurement, and not more than 1 inch out of level over the entire bearing surface.

Interior and exterior walls shall not vary more than 1/4 inch out of square when measured along the diagonal of a 6x8x10-foot triangle at any location. Walls shall not be more than 1/8 inch out of plumb in any 48-inch vertical measurement.

Roof ridge beam deflection shall be no greater than 1 inch in 8 feet.

Roof trusses shall be installed per the manufacturer's layout. Roof members should be installed securely and be within 1/2 inch of plumb from collar tie to top of ridge. Ridge shall not vary more than 1/2 inch out of level plus or minus over the entire length.

If doors and windows are installed by the framing Trade Contractor, doors and windows shall be installed level, plumb in both directions, and squarely into the opening with no more than a 1/4-inch-in-4-feet deviation in any direction. Windows and doors should be installed so that they operate properly as intended and with reasonable ease.

Materials: All framing materials should be of appropriate size, species, and grade to meet building code requirements and/or specifications where they exceed code requirements. Lumber that is inferior should be used in noncritical and/or nonload-bearing areas. When the effect of a crooked or structurally inferior piece of lumber cannot be minimized it should not be installed.

All bottom plates must be pressure-treated material.

Installation: All work is to be done by trained, experienced individuals. Notching, drilling, and cutting of framing members shall be done so as to not to compromise the structural integrity of the house.

Plywood subfloor panels shall be installed with grain direction perpendicular to joists, and with staggered end joints. Adjacent panels shall be spaced 1/32 inch apart, with those intersecting vertical surfaces 1/8 inch apart. Floor sheathing shall be level and securely attached to supports to provide an adequate base to receive finish flooring and fasteners.

Studding and other supports shall be adequately sized and spaced. Wall sheathing shall be smooth, securely attached to supports, and shall provide an adequate base to receive siding and fasteners.

Roof sheathing shall be smooth, securely attached to supports, and shall provide an adequate base to receive roofing nails and fasteners.

Window and door headers shall conform to code requirements and be installed crown up. Window and doors should be checked with a level for jamb squareness and straightness.

_____ Company Rep's Initials
_____ Trade Contractor's Initials

Other: The framing Trade Contractor is responsible for ensuring that the house is framed per plan with all dimensions correct, work plumb, square, and of the highest quality. The house's final appearance begins with the framing Trade Contractor. The quality of the work of the Trade Contractors following the framing Trade Contractor depends in large part on the quality of the framing Trade Contractor's work.

The Company produces its own plans (for the most part) and must rely on the Trade Contractor to note any incorrect items found during framing. Should an error be found please notify the Site Superintendent immediately so that corrections may be made.

Floor and/or roof trusses are used in all homes. All floor trusses are clearly marked with a stamp saying, "Top." Bearing tags are attached to trusses marking interior bearing locations. It is the responsibility of the Trade Contractor to install all trusses correctly according to the layout furnished with the trusses and to plan.

The Trade Contractor shall repair any subflooring that squeaks, creaks, or gives due to improper installation. Any buckling of the subfloor caused by the subflooring being installed too close together shall be planed, sanded, and smoothed by the framer. Any gaps larger than 1/8 inch in the subflooring shall be filled with durorock and sanded smooth.

Any roof decking that is warped, buckled, or loose will be replaced or repaired by the Trade Contractor. Roof decking shall not bow more than 1/2 inch in 2 feet.

Warranty: The Company believes that all work done in connection with one of our homes should be of high quality and that all Trade Contractors should stand behind the quality of their work. Therefore we require all Trade Contractors to warrant the quality of their work for a period of one (1) year from date of closing of the house. Please refer to our printed Limited Warranty booklet for specific items that are covered under Warranty as they apply to the framing.

The Trade Contractor shall have seven (7) days in which to correct any Warranty problem. If the problem is not corrected within seven (7) days then The Company shall correct the problem and will backcharge the Trade Contractor at the rate of $25.00 per hour, with a minimum charge of $100.00 plus the cost of any materials.

Inspection Reports: The Trade Contractor and the Site Superintendent shall walk the job together and complete each section of the inspection report(s). The Trade Contractor must correct any deficiency found during the inspection and the job must be 100-percent complete before payment will be made. The Trade Contractor and the Site Superintendent must sign-off on all sections of the inspection report(s) attesting that the job is 100-percent complete and is correct per the job requirements found in this Scope of Work.

Detailed Job Procedures:

1. A new set of plans is required for each house. Plans are subject to changes and modifications. It is the responsibility of the Trade Contractor to have the new plans before beginning work. Plans should be picked up at the job trailer from the Site Superintendent. The Trade Contractor at no cost to The Company will correct any errors that occur from using an incorrect set of plans.

Company Rep's Initials _____
Trade Contractor's Initials _____

2. Purchase orders should be picked up from the Site Superintendent.

3. The Trade Contractor and the Site Superintendent must walk the job together and complete the pre-work section of the inspection report(s) before work may begin. The pre-work section of the inspection report(s) must be signed-off on by both parties.

4. The Trade Contractor will be furnished with a copy of the material order for the job. If any items are missing or substituted the Site Superintendent should be informed immediately.

5. The Trade Contractor shall keep the waste of materials to a minimum. Wasted or damaged materials will be at the expense of the Trade Contractor.

6. Sill sealer insulation shall be installed in all exterior walls and walls separating the house from the garage. Sill sealer shall extend slightly on each side of a 2x4 plate. It must be visible for all inspections.

7. All bottom plates shall be 2x4 pressure-treated wood. In no circumstances shall material that is not pressure treated come in contact with concrete.

8. The foundation Trade Contractor is responsible for setting anchor bolts into the foundation. The framing Trade Contractor is responsible for boring the necessary holes to accommodate the anchor bolts. The framing Trade Contractor is responsible for the completion of the anchor bolt installation with washers and nuts.

9. All exterior walls shall be constructed of 2x4 studs set 16 inches on center.

10. Interior nonload-bearing walls shall be constructed of 2x4 studs set 24 inches on center.

11. Interior load-bearing walls shall be constructed of 2x4 studs set 16 inches on center.

12. If the plans call for 2x6 studs for interior load-bearing walls they shall be 16 inches on center.

13. Basement load-bearing walls shall be 2x6 studs set 16 inches on center.

14. All top plates shall be doubled 2x4 utility-grade material.

15. All headers shall be true and level using 2x10 material.

16. All triple windows shall use laminated beams in place of 2x10 material.

17. All window and door openings shall be level, plumb, square, and built to the correct rough opening size as specified by the manufacturer.

18. Individual double garage doors shall have a laminated beam installed as the header. Individual single doors shall have a 2x10 #1 material installed for headers. Headers must be completely and securely resting on, and fastened to, supporting jacks. If the plans call for brick, the brick lentil must be installed so that it reaches the outside edge of the supporting 2x4 jacks.

19. All bifold doors over carpet with jambs and casing shall be 82 inches high and 2 inches wider than the doors.

_____ Company Rep's Initials
_____ Trade Contractor's Initials

20. All bifold doors over vinyl with jambs and casings shall be 81 inches high and 2 inches wider than the doors.

21. Bifold doors with drywall jambs over vinyl shall be 80-1/2 inches high and 1-1/4 inches wider than the doors.

22. Bifold doors with drywall jambs over carpet shall be 81-1/2 inches high and 1-1/2 inches wider than the doors.

23. Exterior and interior doors over carpet shall be 83 inches high and 2 inches wider than the door.

24. Exterior and interior doors over vinyl shall be 82 inches high and 2 inches wider than the door.

25. Windows are purchased with wood surrounds. Windows shall be framed 1-1/2 inches wider and higher than the window size or per the manufacturer's rough opening specifications.

26. Sliding glass doors are purchased with wood surrounds. Sliding doors shall be framed 82 inches high and 1-1/2 inches wider than the door or per the manufacturer's rough opening specifications.

27. Trusses shall be set as per layout from truss manufacturer paying particular attention to "Top" stamp and bearing tags. All joist hangers shall be installed and nailed with a nail in each hanger hole. **No truss may be cut, notched, nicked, or otherwise damaged**. If any truss is accidentally damaged the Site Superintendent should be notified immediately. No truss may be repaired without the approval of the manufacturer's engineer.

28. All exterior corner walls shall be braced using metal lateral bracing.

29. All knee walls shall be constructed in such a manner that there is no give, sway, or movement.

30. All plywood subflooring shall be glued and nailed immediately after the application of the adhesive.

31. There shall be no give, creaking, or squeaking due to loose, improperly installed subfloors.

32. No subflooring shall float at joints or squeak overall.

33. All subflooring shall be installed with a 1/8-inch gap between pieces to allow for swelling due to the absorption of moisture. Gaps larger than 1/8 inch shall be filled with durorock compound, then smoothed and sanded.

34. All subflooring shall fit within 1/16 inch of all walls. Gaps larger than 1/16 inch shall be filled with durorock compound, then smoothed and sanded.

35. Any overlapping of subflooring pieces due to swelling or buckling shall be planed, smoothed, and leveled.

36. The Trade Contractor is responsible for setting temporary stairs before beginning the next level.

37. Temporary handrails shall be installed on all stairs.

Company Rep's Initials _____
Trade Contractor's Initials _____

38. Safety bracing shall be installed around all openings with drops of more than 2 feet. All openings for exterior doors, sliding doors, etc., that have a drop of more than 2 feet shall be blocked with a brace made into an X with a crossbar running horizontally across the opening.

39. Garden tub and tub/shower units shall be placed inside the bath areas before the bathrooms are framed. All framing for bathrooms shall be completed only **after** the tub units are inside.

40. All nails are to be firmly set. No bent nails shall protrude into or onto any area that will receive door casings, drywall, or other types of finish.

41. No warped studs shall be used if the use of such studs will result in a wall that is out of plumb more than 3/8 of an inch in any 32 inches in any 8-foot vertical measurement or if the use will result in a wall that bows more than 1/2 inch out of line within any 32-inch horizontal measurement, or 1/2 inch within any 8-foot vertical measurement.

42. All firestops shall be installed per code (chimney chases, under tubs, around showers, etc.).

43. Adequate deadwood at shower stalls, windows, doors, corners, and other areas, which ensures an adequate solid surface area for doors or exterior trim anchorage, shall be installed by the Trade Contractor.

44. Decking shall be installed with plyclips set 24 inches on center.

45. Decking shall be securely attached to supports and shall provide an adequate base to receive roofing nails and fasteners. Decking shall not bow more than 1/2 inch in 2 feet.

46. Holes shall be cut per plans for all roof vents.

47. Exterior sheathing shall be installed according to manufacturer's instructions. No broken pieces of sheathing shall be left in place. No gaps or missing pieces shall be allowed.

48. Gable ends shall be completely sheathed.

49. Wind bracing shall be installed at gable ends.

50. Scuttle holes or disappearing stairs shall be framed according to plan.

51. In all houses where the plans call for two heating and air conditioning (HVAC) units or where the plans call for the HVAC unit to be placed in the attic, the Trade Contractor must frame a minimum 20-inch-wide-by-30-inch-long access opening for the unit(s) in the attic. A passageway to the area of the HVAC unit must be a minimum of 22 inches wide by 30 inches high. The passageway must have a minimum 22-inch-wide floor. Flooring shall extend a minimum of 30 inches in width along the control side of the equipment with a 30-inch-high clear working space on all sides where access is necessary for servicing the HVAC unit. If two units are to be installed in the attic the flooring area must be increased to accommodate two units.

52. If the house plans call for the hot water tank to be located in the garage area rather than in a storage room in the garage, the Trade Contractor is required to frame two walls, with the third

_____ Company Rep's Initials
_____ Trade Contractor's Initials

wall being the house wall, in the area of the hot water tank. One wall must protect the hot water tank from impact or damage by an automobile.

53. If the Trade Contractor is required to set windows and exterior doors the Trade Contractor shall follow and abide by the Window and Door Installation Scope of Work. Door and window installation will be checked using the Window and Door Installation inspection report.

54. Deadwood shall be installed throughout the house to ensure proper installation of drywall.

55. Thresholds shall be braced before the job is complete.

56. All temporary bracing, except safety bracing, shall be removed at the completion of the framing.

57. All trash and building debris shall be removed to the dumpster or to an area designated by the Site Superintendent.

58. The house shall be broom-swept before job shall be considered complete.

59. All extra material will be stacked in the garage and the Site Superintendent together with the Trade Contractor will inventory the excess material. The inventory list shall be attached to the purchase order.

60. The house must pass all framing inspections.

61. Any corrections required by the framing inspectors will be at the expense of the Trade Contractor.

62. The Trade Contractor and Site Superintendent must walk the job together and perform a final inspection of the job. The final section of the inspection report(s) must be completed and signed-off on by both parties. The inspection report(s) must be attached to the office's copy of the purchase order and the Trade Contractor's invoice or payment will not be issued.

63. Any items found during the final inspection that need correction shall be corrected before payment will be made.

I _____ agent for _____

_____ have read and fully understand the above **Scope of Work** and I hereby agree to perform all work in accordance with the above.

Date: _____   _____
                            Signed: Trade Contractor (or agent)

Date: _____   _____
                            For The Company

INSPECTION REPORT

FRAMING LABOR

Subdivision/Lot # _____

Pre-Work Inspection: (to be completed prior to beginning work)

____ Plans have been picked up and signed for.

____ Purchase order has been picked up.

____ Quantity of materials ordered list has been received.

____ Work area is clean and free of debris.

____ Foundation is complete, square, plumb, and anchor bolts are in place.

____ All formboards have been removed from foundation.

____ Material is on site and ready to be used.

____ Access for crane is adequate.

____ Trusses are on site and ready to install.

____ Temporary electrical power is available.

Date: _____   Site Superintendent _____

Trade Contractor _____

Final Inspection: (to be completed before Trade Contractor leaves jobsite)

____ Sill sealer insulation has been installed at all exterior walls and walls separating house from garage. Sill sealer extends slightly on each side of 2x4 plates. It is visible for all inspections.

____ All bottom plates are 2x4 pressure-treated wood. No nonpressure-treated material is in contact with concrete.

____ All anchor bolts have washers and nuts.

____ All exterior walls are constructed of 2x4 studs set 16 inches on center.

____ Interior nonload-bearing walls are constructed of 2x4 studs set 24 inches on center.

____ Interior load-bearing walls are constructed of 2x4 studs set 16 inches on center.

____ If the plans call for 2x6 studs in interior load-bearing walls they are set 16 inches on center.

____ Basement load-bearing walls are 2x6 studs set 16 inches on center.

____ All top plates are doubled 2x4 utility-grade material.

____ All headers are true and level using 2x10 material.

____ Laminated beams are in place and secure for individual double garage doors.

____ Headers of 2x10 material, resting securely on jacks, are installed for individual double garage doors.

____ Brick lintel rests securely on and overlaps outside of jacks over garage doors.

____ All triple windows have laminated beams installed in place of 2x10 material.

____ All window and door openings are level, plumb, and square, and built to the correct size. All windows and doors have been checked using a 3-foot level and framing square.

____ All bifold doors over carpet with jambs and casing are 82 inches high and 2 inches wider than the doors or framed per the manufacturer's rough opening specifications.

____ All bifold doors over vinyl with jambs and casings are 81 inches high and 2 inches wider than the doors or framed per the manufacturer's rough opening specifications.

____ Bifold doors with drywall jambs over vinyl are 80-1/2 inches high and 1-1/4 inches wider than the doors or framed per the manufacturer's rough opening specifications.

____ Bifold doors with drywall jambs over carpet are 81-1/2 inches high and 1-1/2 inches wider than the doors or framed per the manufacturer's rough opening specifications.

____ Exterior and interior doors over carpet are 83 inches high and 2 inches wider than the door or framed per the manufacturer's rough opening specifications.

____ Exterior and interior doors over vinyl are 82 inches high and 2 inches wider than the door or framed per the manufacturer's rough opening specifications.

____ All windows are framed 1-1/2 inches wider and higher than the window size or framed per the manufacturer's rough opening specifications.

____ Sliding doors are framed 82 inches high and 1-1/2 inches wider than the door or framed per the manufacturer's rough opening specifications.

____ Trusses are set as per layout from truss manufacturer.

____ All joist hangers are installed and nailed. There are no empty holes.

____ **No truss is cut, notched, nicked, or otherwise damaged.**

____ All exterior corner walls are braced using metal lateral bracing.

____ All knee walls are secure with no give, sway, or movement.

____ All subflooring is glued and nailed.

____ Subfloor does not give, creak, or squeak due to loose, or improperly installed subflooring.

____ No areas of the subfloor float at joints and there are no squeaks.

____ All subflooring is installed with a 1/8-inch gap between pieces.

____ All subflooring fits within 1/16 inch of all walls.

____ Any subfloors with gaps larger than 1/8 of an inch are filled with durorock compound, then smoothed and sanded.

____ Any place with subflooring pieces overlap has been planed, smoothed, and leveled.

____ Temporary stairs are set at each level.

____ Temporary handrails are installed on all stairs.

____ Safety bracing is installed around all openings with drops of more than 2 feet. Such openings are blocked with an "X" with a crossbar running horizontally across the opening.

____ Garden tub and tub/shower units are in set inside bathroom areas.

____ All nails are firmly set. No bent nails are protruding into any area that will receive door casings, drywall, or other types of finishes.

____ No studs are warped.

____ All firestops are installed per code (chimney chases, under tubs, around showers, etc.).

____ Extra blocking is installed at shower stalls, windows, doors, corners, and other areas.

____ Decking is installed with plyclips set 24 inches on center.

____ Decking is level to 1/2 inch in 2 feet.

____ Decking is securely attached to supports in order to receive roofing nails and fasteners.

____ All holes are cut, per plans, for all roof vents.

____ Exterior sheathing is installed according to manufacturer's instructions. There are no broken pieces of sheathing, and no gaps or missing pieces.

____ Gable ends are completely sheathed.

____ Wind bracing is installed at gable ends.

____ Scuttle holes or disappearing stairs are framed according to plans.

____ If framer is required to set windows, all windows are set level, plumb, and square. All windows operate correctly and lock smoothly.

____ If framer is required to set exterior doors, all doors are square, plumb, shimmed, and the required special bolts furnished by the manufacturer are installed per manufacturer's instructions in each door.

____ Deadwood is installed throughout house.

____ Door thresholds are braced with pressure-treated wood.

____ All temporary bracing, except safety bracing, has been removed.

____ All brick lentils are installed as required and reach outside edge of supporting jacks.

____ All trash and building debris has been removed from house and site to correct area.

____ House has been broom-swept.

____ Excess material is in garage.

____ Excess materials have been counted and noted on quantity list.

Date: _____   Site Superintendent _____

                              Trade Contractor _____

SCOPE OF WORK
Standards and Description of Work Performance

GARAGE DOORS

The Company's **Terms and Conditions** are by reference a part of all **Scope of Work** requirements.

Construction Requirements:

Generally speaking the work of Trade Contractors and their employees is expected to be performed in a good and workmanlike manner. Workmanlike quality is defined as workmanship that meets or betters those criteria indicated in applicable building codes, using materials and installation methods identified in the construction plans and this Scope of Work.

Code Requirements:

All jobs shall conform to those standards stipulated in the building code, mechanical code, plumbing code, and electrical code applicable in the local jurisdiction. All construction on The Company's jobsites shall meet or exceed NAHB Performance and Building Standards.

General Comments:

The Company considers our Trade Contractors to be experts at producing a high quality job. But everyone on our construction team—staff, Trade Contractors, and suppliers—must recognize the importance of providing quality in both the product and service areas while on our jobsites and in the homes of our purchasers.

Since we work as a team, poor quality or service, from any of us, reflects unfavorably on all of us. An exceptional level of product quality and highly effective service can help us all to increase our business and grow.

The Company's definition of quality construction also requires that every job be completed correctly the first time. When this does not occur it costs both of us additional money, imposes on the purchaser, and hurts our reputations as quality builders. That is why, in situations where construction was not completed in a quality manner, prompt corrective action is required to remedy specific deficiencies.

In the following information the term Site Superintendent shall refer to any The Company representative with authority to perform the specified task. The term Trade Contractor shall mean the Trade Contractor's organization or any representative that is assigned the authority to perform the specified task.

General: Garage doors are installed in all of The Company's houses. Doors may be single or double depending on the plan.

Company Rep's Initials _____
Trade Contractor's Initials _____

Material: The type, size, make, and grade of garage doors is dependent on the subdivision and plan. Once a type, grade, and make of garage door is selected by The Company it may not be changed without the written permission of The Company's purchasing agent. All doors shall have locking devices installed.

Installation: All installations shall be by trained, experienced personnel. Doors shall be installed level, plumb, and square. They shall be installed per the manufacturer's installation instructions and shall operate properly and smoothly.

All springs shall be adjusted to ensure doors operate smoothly and correctly. Door locks shall operate smoothly, correctly, and secure door. Weathersealer/weatherstripping shall be installed per manufacturer's instructions.

If garage door opener(s) are installed they shall be installed per the manufacturer's installation instructions and operate smoothly.

Warranty: The Company believes that all work done and all materials installed in connection with one of our homes should be of high quality and that all Trade Contractors should stand behind the quality of their work and materials. Therefore we require all Trade Contractors to warrant the quality of their work for a period of one (1) year from date of closing of the house. Please refer to our printed Limited Warranty booklet for any specific items that are covered under Warranty as they apply to garage doors.

The Trade Contractor shall have seven (7) days in which to correct any Warranty problem. If the problem is not corrected within seven (7) days then The Company shall correct the problem and will backcharge the Trade Contractor at the rate of $25.00 per hour, with a minimum charge of $100.00 plus the cost of any materials.

Inspection Reports: The Trade Contractor and the Site Superintendent shall walk the job together and complete each section of the inspection report(s). The Trade Contractor must correct any deficiency found during the inspection and the job must be 100-percent complete before payment will be made. The Trade Contractor and the Site Superintendent must sign-off on all sections of the inspection report(s) attesting that the job is 100-percent complete and is correct per the job requirements found in this Scope of Work.

Detailed Job Requirements:

1. The Trade Contractor and the Site Superintendent must walk the job together and complete the pre-work section of the inspection report(s) before work may begin. The pre-work section of the inspection report(s) must be signed-off on by both parties.

2. Purchase orders will be mailed to the Trade Contractor from the office.

3. Garage doors shall be of a type, make, and grade specified by The Company. No substitution is allowed.

4. All garage doors shall be installed per the manufacturer's installation instructions.

_____ Company Rep's Initials
_____ Trade Contractor's Initials

5. Doors shall be installed level, plumb, square, and scribed to the existing garage concrete floor.

6. Doors shall operate smoothly and correctly.

7. Installation shall be such as to not damage existing work and drywall.

8. Doors shall be undamaged.

9. Locking devices shall operate correctly and cause the door to be secure when locked.

10. Garage door key(s) will be left on a nail in the wall of the garage to the left of a single double-door or between two single doors.

11. All doors must be weathersealed per manufacturer's instructions.

12. If garage door opener(s) is included, the opener(s) shall be installed per the manufacturer's installation instructions. When opener is used the garage door must open smoothly with no binding or jamming.

13. All construction debris must be removed to the dumpster or to an area designated by the Site Superintendent. The job will not be considered complete and payment will not be issued until all trash and debris have been removed from the house and site.

14. Garage shall be left clean and broom-swept.

15. The Trade Contractor and Site Superintendent must walk the job together and perform a final inspection of the job. The final section of the inspection report(s) must be completed and signed-off on by both parties. The inspection report(s) must be attached to the office's copy of the purchase order and the Trade Contractor's invoice or payment will not be issued.

16. Correction of any items found on the final inspection report must be completed prior to payment being issued.

I _____ agent for _____

_____ have read and fully understand the above **Scope of Work** and I hereby agree to perform all work in accordance with the above.

Date: _____   _____
                            Signed: Trade Contractor (or agent)

Date: _____   _____
                            For The Company

INSPECTION REPORT

GARAGE DOORS

Subdivision/Lot # _____

Pre-Work Inspection: (to be completed prior to beginning work)

____   Plans have been reviewed with Site Superintendent.

____   Purchase order has been received.

____   Garage door opening(s) is framed and the dimensions verified as correct.

____   Headers/lintels, etc., are installed level, plumb, and square.

____   Garage is clean, broom-swept, and free of debris.

Date: _____   Site Superintendent _____

                                             Trade Contractor _____

Final Inspection: (to be completed before Trade Contractor leaves jobsite)

____   Garage door(s) is installed using correct number, size, type, color, etc.

____   Door(s) operates smoothly.

____   Door lock(s) operates smoothly and correctly.

____   Door keys are hung on wall in proper location.

____   Thresholds are correct.

____   Weathersealer/weatherstripping is installed.

____   Spring tension is adjusted properly.

____   Door(s) is level, plumb, square, and scribed to existing garage floor.

____   Door(s) is undamaged.

____   Garage is clean, broom-swept, and free of debris.

____   Construction debris has been removed to designated location.

____   Door opener(s) operate correctly. Openers are hung in accordance with manufacturer's specifications. One opener per door has been received.

Date: _____   Site Superintendent _____

                                             Trade Contractor _____

<u>SCOPE OF WORK</u>
Standards and Description of Work Performance

GUTTERING

The Company's **<u>Terms and Conditions</u>** are by reference a part of all **<u>Scope of Work</u>** requirements.

<u>Construction Requirements</u>:

Generally speaking the work of Trade Contractors and their employees is expected to be performed in a good and workmanlike manner. Workmanlike quality is defined as workmanship that meets or betters those criteria indicated in applicable building codes, using materials and installation methods identified in the construction plans and this Scope of Work.

<u>Code Requirements</u>:

All jobs shall conform to those standards stipulated in the building code, mechanical code, plumbing code, and electrical code applicable in the local jurisdiction. All construction on The Company's jobsites shall meet or exceed NAHB Performance and Building Standards.

<u>General Comments</u>:

The Company considers our Trade Contractors to be experts at producing a high-quality job. But everyone on our construction team—staff, Trade Contractors, and suppliers—must recognize the importance of providing quality in both the product and service areas while on our jobsites and in the homes of our purchasers.

Since we work as a team, poor quality or service, from any of us, reflects unfavorably on all of us. An exceptional level of product quality and highly effective service can help us all to increase our business and grow.

The Company's definition of quality construction also requires that every job be completed correctly the first time. When this does not occur it costs both of us additional money, imposes on the purchaser, and hurts our reputations as quality builders. That is why, in situations where construction was not completed in a quality manner, prompt corrective action is required to remedy specific deficiencies.

In the following information the term Site Superintendent shall refer to any The Company representative with authority to perform the specified task. The term Trade Contractor shall mean the Trade Contractor's organization or any representative that is assigned the authority to perform the specified task.

<u>General:</u> The Trade Contractor is to furnish all materials and labor to install guttering per plan. Material is to be specified at the bid stage for each model per subdivision.

<u>Installation:</u> All installations are to be by trained, experienced workers. Installations shall be per the manufacturer's installation instructions using the fasteners and procedures specified.

Company Rep's Initials _____
Trade Contractor's Initials _____

Safety: All safety precautions must be observed at all times. All ladders, scaffolding, bracing, etc., shall be per OSHA requirements.

Warranty: The Company believes that all work done and all materials installed, in connection with one of our homes should be of high quality and that all Trade Contractors should stand behind the quality of their work and materials. Therefore we require all Trade Contractors to warrant the quality of their work for a period of one (1) year from date of closing of the house. Please refer to our printed Limited Warranty booklet for any specific items that are covered under Warranty as they apply to the guttering.

The Trade Contractor shall have seven (7) days in which to correct any Warranty problem. If the problem is not corrected within seven (7) days then The Company shall correct the problem and will backcharge the Trade Contractor at the rate of $25.00 per hour, with a minimum charge of $100.00 plus the cost of any materials.

Inspection Reports: The Trade Contractor and the Site Superintendent shall walk the job together and complete each section of the inspection report(s). The Trade Contractor must correct any deficiency found during the inspection and the job must be 100-percent complete before payment will be made. The Trade Contractor and the Site Superintendent must sign-off on all sections of the inspection report(s) attesting that the job is 100-percent complete and is correct per the job requirements found in this Scope of Work.

Detailed Job Requirements:

1. Plans shall be reviewed with the Site Superintendent prior to beginning work.

2. Purchase order will be mailed to the Trade Contractor from the office.

3. The Trade Contractor and the Site Superintendent must walk the job together and complete the pre-work section of the inspection report(s) before work may begin. The pre-work section of the inspection report(s) must be signed-off on by both parties.

4. Material and type of gutters shall be the size and type specified on the purchase order.

5. All gutters are to be firmly attached to the house with a minimum of gaps between roof and gutter.

6. All gutters shall be installed with the correct slope for drainage. Water shall not stand in gutters more than 1/2 inch in any area.

7. All downspouts shall be firmly attached to the house with adequate straps to ensure there are no bows or gaps between wall and downspout.

8. Downspouts should be located approximately every 30 feet or as required by roof design.

9. Downspouts shall terminate no higher than 6 inches from grade.

10. Splashguards shall be installed as necessary.

_____ Company Rep's Initials
_____ Trade Contractor's Initials

11. Splashblocks shall be installed at each downspout if landscaping is complete. If landscaping is not complete splashblocks shall be left in garage for installation after landscaping is complete.

12. Downspouts shall not terminate in such a manner as to cause water-flow problems. The Site Superintendent shall have the final authority to change the direction of the downspout water flow.

13. Downspouts shall be continuous as much as possible with a minimum of piecing and joining.

14. No downspout may terminate on a deck. A hole shall be cut in the deck and the downspout is to continue through the deck to 6 inches above grade.

15. Splashguards shall be installed at all valleys.

16. Gutters shall not leak at miters, corners, or joints.

17. If sod has been laid either a piece of sod shall be removed (and replaced) or ladders shall be placed on boards to prevent damage to sod.

18. All debris shall be removed to the dumpster or designated trash area and the jobsite shall be left clean and free of debris.

19. The Trade Contractor is responsible for the cleaning up of all residual materials before the job will be accepted as complete.

20. The Trade Contractor and Site Superintendent must walk the job together and perform a final inspection of the job. The final section of the inspection report(s) must be completed and signed-off on by both parties. The inspection report(s) must be attached to the office's copy of the purchase order and the Trade Contractor's invoice or payment will not be issued.

I _____ agent for _____

_____ have read and fully understand the above **Scope of Work** and I hereby agree to perform all work in accordance with the above.

Date: _____

_____
Signed: Trade Contractor (or agent)

Date: _____

_____
Signed: For The Company

INSPECTION REPORT

GUTTERING

Subdivision/Lot # _____

Pre-Work Inspection: (to be completed prior to beginning work)

____   Plans have been reviewed, the house walked, and any special areas noted.

____   Purchase order has been received.

____   Outside area of house is clean and free of construction debris.

____   Siding is complete. All trim is installed and undamaged.

____   Roofing is complete.

____   Final grading is complete.

____   Windows are intact and none are broken.

____   If yard is sodded, sod is undamaged.

Date: _____   Site Superintendent _____

                              Trade Contractor _____

Final Inspection: (to be completed before Trade Contractor leaves jobsite)

____   Windows are intact and none are broken.

____   Trim and siding are undamaged.

____   Sod is undamaged.

____   Gutters are firmly attached to house.

____   Gutters and downspouts contain only minor piecing.

____   Gutters have no holes or gaps.

____   An adequate number of downspouts are installed.

____   Downspouts are firmly attached to house with no gaps or bows.

____   Downspouts terminate no more than 6 inches from grade.

____   Downspouts continue through deck if necessary.

____   Downspout hole through deck is cut smoothly and to the correct size.

____   Splashblocks are in place or are in garage.

____   Water flow from downspouts is correct.

____   Splashguards are in place in valleys.

___   Site is clean and free of debris.

Date: _____   Site Superintendent _____

                                          Trade Contractor _____

<u>SCOPE OF WORK</u>
Standards and Description of Work Performance

HEATING and AIR CONDITIONING

The Company's **Terms and Conditions** are by reference a part of all **Scope of Work** requirements.

Construction Requirements:

Generally speaking the work of Trade Contractors and their employees is expected to be performed in a good and workmanlike manner. Workmanlike quality is defined as workmanship that meets or betters those criteria indicated in applicable building codes, using materials and installation methods identified in the construction plans and this Scope of Work.

Code Requirements:

All jobs shall conform to those standards stipulated in the building code, mechanical code, plumbing code, and electrical code applicable in the local jurisdiction. All construction on The Company's jobsites shall meet or exceed NAHB Performance and Building Standards.

General Comments:

The Company considers our Trade Contractors to be experts at producing a high-quality job. But everyone on our construction team—staff, Trade Contractors, and suppliers—must recognize the importance of providing quality in both the product and service areas while on our jobsites and in the homes of our purchasers.

Since we work as a team, poor quality or service, from any of us, reflects unfavorably on all of us. An exceptional level of product quality and highly effective service can help us all to increase our business and grow.

The Company's definition of quality construction also requires that every job be completed correctly the first time. When this does not occur it costs both of us additional money, imposes on the purchaser, and hurts our reputations as quality builders. That is why, in situations where construction was not completed in a quality manner, prompt corrective action is required to remedy specific deficiencies.

In the following information the term Site Superintendent shall refer to any The Company representative with authority to perform the specified task. The term Trade Contractor shall mean the Trade Contractor's organization or any representative that is assigned the authority to perform the specified task.

General: The heating system should be a complete and functioning system tested and ready for operation. The equipment and system shall be installed in accordance with all approved building, plumbing, electrical, and health codes and function as specified. Installations shall pass the initial compliance inspection at each phase.

_____ Company Rep's Initials
_____ Trade Contractor's Initials

The heating and cooling systems should operate correctly for the first two (2) years of Warranty coverage with the exception of those items caused by owner negligence.

The heating system should be capable of producing an inside temperature of 70 degrees Fahrenheit (F), as measured in the center of each room at a height of 5 feet above the floor, under local outdoor winter design conditions as specified in the ASHRAE handbook. Federal, state, or local energy codes shall supersede this standard where such codes have been adopted locally.

The cooling system shall be a complete and functioning system tested and ready for operation. The equipment and system shall be installed in accordance with all approved building, plumbing, electrical, and health codes, and function as specified. Installations should pass the initial code inspection at each phase.

The cooling system shall be capable of maintaining a temperature of 78 degrees F, as measured in the center of each room at a height of 5 feet above the floor, under local outdoor summer design conditions as specified in the ASHRAE handbook. In the case of outside temperatures exceeding 95 degrees F, a differential of 15 degrees F from the outside temperature shall be maintained. Federal, state, or local energy codes shall supersede this standard where such codes have been adopted locally.

The stiffening of ductwork and the gauge of the metal used shall be such that ducts do not "oil can." Refrigerant lines should not develop leaks and there should be no condensation on refrigerant lines during the first year of the Warranty period.

The Company will furnish the Trade Contractor a full set of plans in a timely manner to allow the Trade Contractor time to complete the necessary heat-load calculations and HVAC layouts. It is the responsibility of the Trade Contractor to determine the tonnage and number of units required for each house.

Materials: Systems shall be installed using new materials of the grade and quality required to meet or exceed local or state and ASHRAE standards. The type and manufacturer of the heating and cooling system shall be approved by The Company and cannot be changed without the written approval of The Company's purchasing agent. Approval must be received in writing. All systems must be energy efficient per current code. Should the requirements for HVAC systems change because of any change in the applicable codes, the Trade Contractor must notify The Company immediately and all plans must be rebid.

Installation: All work is to be done by trained, experienced individuals. Workmanship shall be so as to require a minimum of repairs and patching after installation. **No roof or floor truss may be cut, notched, moved, or otherwise damaged in any manner**. If any truss interferes with the installation of the HVAC system, the Site Superintendent should be notified so that the truss manufacturer may solve the problem. Should any truss be damaged accidentally the Site Superintendent must be notified immediately. The truss manufacturer's engineer must approve any repairs before they are made.

Warranty: Certain items are warrantable for one (1) year and some for two (2) years. Refer to The Company's printed Limited Warranty booklet for a list of these items. The Trade Contractor will furnish to the homeowner, via a tag on the HVAC unit and on the sticker on the inside of the left door of the

Company Rep's Initials _____
Trade Contractor's Initials _____

cabinet located to the left of the kitchen sink, a regular office hour phone number and an emergency phone number. The Trade Contractor is required to furnish emergency service.

The Trade Contractor shall have seven (7) days in which to correct any Warranty problem. If the problem is not corrected within seven (7) days then The Company shall correct the problem and will backcharge the Trade Contractor at the rate of $25.00 per hour, with a minimum charge of $100.00 plus the cost of any materials.

Inspection Reports: The Trade Contractor and the Site Superintendent shall walk the job together and complete each section of the inspection report(s). The Trade Contractor must correct any deficiency found during the inspection and the job must be 100-percent complete before payment will be made. The Trade Contractor and the Site Superintendent must sign-off on all sections of the inspection report(s) attesting that the job is 100-percent complete and is correct per the job requirements found in this Scope of Work.

Detailed Job Procedures:

1. A new set of plans is required for each house. Plans are subject to changes and modifications. It is the responsibility of the Trade Contractor to have the new plans before beginning work. Plans should be picked up at the job trailer from the Site Superintendent. The Trade Contractor at no cost to The Company will correct any errors that occur from using an incorrect set of plans.

2. Purchase orders will be mailed from the office. If a purchase order has not been received prior to the beginning of the job one should be picked up from the Site Superintendent.

3. The Trade Contractor and the Site Superintendent must walk the job together and complete the pre-work section of the inspection report(s) before work may begin. The pre-work section of the inspection report(s) must be signed-off on by both parties.

4. The Trade Contractor must be licensed in the county in which the subdivision is located prior to beginning work.

5. The Trade Contractor is responsible for obtaining all necessary permits.

6. Heating and cooling equipment shall be installed per manufacturers' instructions and all applicable codes.

7. Trade Contractor shall be responsible for installing vent for hot water tank.

8. The framing Trade Contractor is responsible for framing and completing the attic access and platform area for all HVAC units to be installed in attic. If the attic area is not complete or completed incorrectly the Site Superintendent should be notified immediately before the unit(s) is installed.

9. The attic unit(s) requires a metal condensation drain pan.

10. All heating and cooling equipment shall be located with sufficient clearances from walls or other equipment to permit cleaning of heating and cooling units, replacement of filters, blowers, motors,

_____ Company Rep's Initials
_____ Trade Contractor's Initials

controls, vent connections, lubrication of moving parts, and adjustments. Should any area be inadequate to allow for the above the Site Superintendent should be notified immediately.

11. The Trade Contractor shall furnish a prebuilt pad for the AC unit. It is to be installed level and square.

12. All ducts should be properly sealed at the heat exchanger per energy code and should not become loose or detached.

13. All ducts shall be installed per the manufacturer's installation instructions. No duct shall be crimped, bent, or otherwise installed in such a manner to restrict airflow. All duct work leaving the heating or cooling unit shall not obstruct the unit in such a manner that the unit cannot be cleaned, filters changed, maintenance performed, etc.

14. **Under no circumstance shall the roof or floor trusses be cut, notched, or otherwise damaged**. Should any truss be accidentally damaged the Site Superintendent should be notified immediately. No repairs may be made without the approval and direction of the manufacturer's engineer.

15. All floor ducts are to be covered to prevent debris from falling into the ductwork.

16. All floor duct openings shall be level, square, and flush with the floor. When grills are installed they should not tilt more than 1/16 inch when stepped on.

17. All ceiling grills shall be square and set flush with the ceiling. They shall have the proper insulation materials installed and shall fit snugly.

18. All gas lines shall be of a material approved by the applicable gas company and code.

19. The Trade Contractor is responsible for the installation of gas lines and the necessary connections for the gas starter in the fireplace, hot water tank, kitchen range, and clothes dryer.

20. The Company uses either concrete siding or vinyl siding. If siding is on the house and lines have not yet been piped to the outside of the house, the Trade Contractor is responsible for **cutting** a neat, clean hole and for completely sealing the hole. Any excess sealing material is to be removed.

21. All gas cutoffs must be properly installed and readily accessible (clothes dryer, fireplace, and furnace).

22. All piping joints and connections shall be suitable for the pressure-temperature conditions and compatible with the piping material.

23. Programmable thermostats shall be installed at location(s) noted on plan. The instruction booklet is to be placed in kitchen cabinet drawer to the left of the range.

24. Condensation pumps shall be plugged in and operational.

25. Heating units shall have filters installed.

26. The vent and grill in the kick plate under the cabinets shall be installed high enough to allow shoe molding to be installed below the kick plate.

Company Rep's Initials _____
Trade Contractor's Initials _____

27. At completion of job house shall be broom-swept and all debris removed to the dumpster or a designated trash area.

28. The Trade Contractor is responsible for passing all required inspections.

29. After gas inspection is completed the gas pressure gauge must be removed and pipe capped. The gas company will not hook up service until gauge is removed and pipe is capped.

30. Any reinspection fees assessed The Company will be charged to the Trade Contractor.

31. The Trade Contractor is responsible for cleaning up all residual materials before the job will be accepted as complete.

32. The Trade Contractor and Site Superintendent must walk the job together and perform a final inspection of the job. The final section of the inspection report(s) must be completed and signed-off on by both parties. The inspection report(s) must be attached to the office's copy of the purchase order and the Trade Contractor's invoice or payment will not be issued.

33. Any items found during the final inspection that need correction shall be corrected before payment will be made

I _____ agent for _____

_____ have read and fully understand the above **Scope of Work** and I hereby agree to perform all work in accordance with the above.

Date: _____   _____
                            Signed: Trade Contractor (or agent)

Date: _____   _____
                            For The Company

INSPECTION REPORT

HEATING and AIR CONDITIONING (ROUGH)

Subdivision/Lot # _____

Pre-Work Inspection: (to be completed prior to beginning work)

____   Plans have been picked up and signed for.

____   Purchase order has been received.

____   Permits have been received.

____   Necessary framing (chases, etc.) are complete.

____   Temporary stairs are in place.

____   Temporary handrails are in place.

____   Safety bracing is in place.

____   Plumber has completed work, pipes are in place, etc.

____   Interior work area is clean and free of debris.

____   Site is clean and free of debris.

____   Electricity is available.

____   If units are to be set in attic, the scuttle hole for attic HVAC equipment access is framed, the HVAC unit area is floored, and adequate working area is available.

____   House is roofed and shingles are installed.

____   HVAC units have been verified for correct type and tonnage.

Date: _____   Site Superintendent _____

                        Trade Contractor _____

Final Inspection: (to be completed before Trade Contractor leaves jobsite)

____   Heating unit(s) has been set with the necessary clearances on all sides.

____   No ductwork is crimped, bent, or damaged so as to restrict airflow.

____   Adequate space in heating unit area has been verified so that ductwork will not obstruct necessary maintenance, changing of filters, etc.

____   Attic unit is properly set, and ducts are not crimped, bent, or otherwise damaged if units are to be placed in attic.

____   All piping to the exterior of the house is complete and sealed properly.

____   All ducts are properly secured to boots.

____ All ducts are properly sealed at the heat exchanger.

____ All ductwork is properly secured.

____ All air returns and vents are properly sealed to floor and placed per plan.

____ All floor ducts are securely covered.

____ All gas cutoffs are installed properly and are accessible (fireplace, clothes dryer, and furnace).

____ Locations of thermostats are correct per plan.

____ Gas pressure is adequate.

____ System has passed rough HVAC inspection.

____ House is broom-swept and clear of debris.

____ No debris is left on lot.

Date: _____   Site Superintendent _____

                           Trade Contractor _____

INSPECTION REPORT

HEATING and AIR CONDITIONING (FINAL)

Subdivision/Lot # _____

Initial Inspection: (to be completed prior to beginning work)

____ Purchase order has been received.

____ Drywall is complete with no damage.

____ Painting is complete with no damage.

____ Trim is complete with no damage.

____ Windows are intact and unbroken.

____ No damage to vinyl, cabinets, etc., is visible.

____ House is clean and free of debris.

Date: _____   Site Superintendent _____

                       Trade Contractor _____

Final Inspection: (to be completed before Trade Contractor leaves jobsite)

____ Coverings are removed from ductwork.

____ Grills are installed square and level. In vinyl areas grills are flush to floor and do not tilt more than 1/4 inch when stepped on.

____ All grills in ceilings are fitted flush with ceiling and have proper sealing material.

____ All other grills fit snugly.

____ HVAC units have been tested to verify correct airflow.

____ Thermostats are square and level.

____ Thermostats have been tested.

____ Instruction and warranty manuals for thermostats, air conditioner, and heater have been placed in kitchen drawer to left of range.

____ Air conditioning condensation unit is set level and operating.

____ Condensation pump, if applicable, is clean, free of debris, and operating properly.

____ For attic unit, condensation pan and drain pipe are free of any obstructions.

____ Hot water tank vent pipe is installed.

____ After gas inspection, the pressure gage have been removed and the pipe capped.

____ House is clean and free of debris.

___   House has passed final HVAC inspection.

Date: _____   Site Superintendent _____

                          Trade Contractor _____

<u>SCOPE OF WORK</u>
Standards and Description of Work Performance

INSULATION

The Company's **Terms and Conditions** are by reference a part of all **Scope of Work** requirements.

Construction Requirements:

Generally speaking the work of Trade Contractors and their employees is expected to be performed in a good and workmanlike manner. Workmanlike quality is defined as workmanship that meets or betters those criteria indicated in applicable building codes, using materials and installation methods identified in the construction plans and this Scope of Work.

Code Requirements:

All jobs shall conform to those standards stipulated in the building code, mechanical code, plumbing code, and electrical code applicable in the local jurisdiction. All construction on The Company's jobsites shall meet or exceed NAHB Performance and Building Standards.

General Comments:

The Company considers our Trade Contractors to be experts at producing a high-quality job. But everyone on our construction team—staff, Trade Contractors, and suppliers—must recognize the importance of providing quality in both the product and service areas while on our jobsites and in the homes of our purchasers.

Since we work as a team, poor quality or service, from any of us, reflects unfavorably on all of us. An exceptional level of product quality and highly effective service can help us all to increase our business and grow.

The Company's definition of quality construction also requires that every job be completed correctly the first time. When this does not occur it costs both of us additional money, imposes on the purchaser, and hurts our reputations as quality builders. That is why, in situations where construction was not completed in a quality manner, prompt corrective action is required to remedy specific deficiencies.

In the following information the term Site Superintendent shall refer to any The Company representative with authority to perform the specified task. The term Trade Contractor shall mean the Trade Contractor's organization or any representative that is assigned the authority to perform the specified task.

General: While insulating the house is not generally complicated, conscientious installation and consistent fastening is essential for good performance. Installation techniques for construction and insulation types may vary somewhat, however there are several basic principles that remain constant.

Company Rep's Initials _____
Trade Contractor's Initials _____

All work is to be done by trained, experienced individuals. Workmanship shall be so as to require a minimum of repairs after installation.

All areas of the walls, floors, and ceilings facing unheated areas shall be insulated. Vapor barriers shall face areas heated in winter. Insulation shall be placed on the outside, or cold side, of pipes and ducts. Wall portions separating heated and unheated areas in houses shall be insulated. Insulation shall be stuffed in openings around ducts, pipes, and wires between heated and unheated areas. All openings in jacks and headers shall be stuffed with insulation to prevent air penetration at these areas.

Batts shall be butted tightly and fastened snugly against framing members. Fasteners shall be installed so as to avoid gaps or "fish mouths." Rips and tears in the vapor barrier shall be repaired. Insulation shall be cut approximately 1 inch larger than nonstandard width spaces. Cut edge of vapor barrier shall be fastened to framing members. Avoid over-compression of insulation material, which reduces the R-value.

Materials: Insulation materials shall meet The Company's specifications for type, brand, and R-value. Materials should be new, dry, undamaged, and stored to remain that way.

Floor Insulation: Unheated floor insulation shall be applied with the vapor barrier up. Floor insulation may be installed with pointed-end wire fasteners or wire lacing.

Wall Insulation: Insulation should be cut to fit snugly in wall areas without being compressed. Insulation cut short should have the gap filled with small pieces. Insulation should be stuffed in narrow areas between framing members and around windows and doors. These areas should be covered with vapor barrier material.

Ceiling Insulation: Ceiling insulation should cover as much of top wallplate as possible. Where eave venting is installed, leave a 1-inch gap between the top of the ceiling insulation and underside of the roof sheathing. Baffleboards shall be installed as required. To avoid fire danger, do not cover recessed lighting fixtures with insulation.

Warranty: The Company believes that all work done and all materials installed in connection with one of our homes should be of high quality and that all Trade Contractors should stand behind the quality of their work and materials. Therefore we require all Trade Contractors to warrant the quality of their work for a period of one (1) year from date of closing of the house. Please refer to our printed Limited Warranty booklet for any specific items that are covered under Warranty as they apply to insulation.

The Trade Contractor shall have seven (7) days in which to correct any Warranty problem. If the problem is not corrected within seven (7) days then The Company shall correct the problem and will backcharge the Trade Contractor at the rate of $25.00 per hour, with a minimum charge of $100.00 plus the cost of any materials.

Inspection Reports: The Trade Contractor and the Site Superintendent shall walk the job together and complete each section of the inspection report(s). The Trade Contractor must correct any deficiency found during the inspection and the job must be 100-percent complete before payment will be made. The Trade Contractor and the Site Superintendent must sign-off on all sections of the

_____ Company Rep's Initials
_____ Trade Contractor's Initials

inspection report(s) attesting that the job is 100-percent complete and is correct per the job requirements found in this Scope of Work.

Detailed Job Requirements:

1. Prior to beginning work the Trade Contractor must review the plans with the Site Superintendent to ensure that the Trade Contractor has a full understanding of the job requirements.

2. The purchase order should be received prior to beginning work. If not, a copy can be picked up from the Site Superintendent.

3. The Trade Contractor and the Site Superintendent must walk the job together and complete the pre-work section of the inspection report(s) before work may begin. The pre-work section of the inspection report(s) must be signed-off on by both parties.

4. All insulation must meet or exceed the R-value specified in the applicable state energy code for single-family homes and FHA standards.

5. All wall insulation shall be 3-1/2-inch wallpaper-backed insulation with an R-13 value including insulation for block foundation walls in basements, if the basement is to be finished.

6. All ceiling insulation will be blown insulation with a R-30 value, except for tray-ceiling and vaulted-ceiling insulation, which shall have R-30 value batts.

7. The Trade Contractor shall furnish one ruler per each 500 square feet of attic space with a minimum of three (3) rulers. Rulers must be readily visible in the attic. The state energy code states that the average R-30 blown insulation should measure a minimum of 9-1/2 inches in depth in all areas of the ceiling.

8. All attic access areas shall have weatherstripping.

9. All areas of penetration in the roof, siding, or floor, such as penetrations for plumbing, wiring, etc., shall be sealed.

10. Areas that cannot be insulated using normal batts, etc., because of size, shape, or installed fixtures or units shall be insulated by stuffing insulation around and in the areas. No area is to be left uninsulated.

11. All areas between walls and roof/ceilings and between panels shall be sealed.

12. All areas behind tubs, showers, etc., must be insulated.

13. Baffelboards shall be installed as required to ensure adequate ventilation in attic area if eve vents are installed in house.

14. All trash and building debris must be removed to the dumpster or to the site designated for trash.

15. House shall be left clean and broom-swept for job to be considered complete.

Company Rep's Initials _____
Trade Contractor's Initials _____

16. The Trade Contractor is responsible for cleaning up all residual materials before the job will be accepted as complete.

17. A letter certifying the R-values of the insulation must be furnished for each house. The letter should be in a format approved by FHA and VA.

18. The Trade Contractor and Site Superintendent must walk the job together and perform a final inspection of the job. The final section of the inspection report(s) must be completed and signed off on. The inspection report(s) must be attached to the office's copy of the purchase order and the Trade Contractor's invoice or payment will not be issued.

19. Any items found during the final inspection that need correction shall be corrected before payment will be made.

I _____ agent for _____

_____ have read and fully understand the above **Scope of Work** and I hereby agree to perform all work in accordance with the above.

Date: _____   _____
                         Signed: Trade Contractor (or agent)

Date: _____   _____
                         For The Company

INSPECTION REPORT

INSULATION

Subdivision/Lot # _____

Pre-Work Inspection: (to be completed prior to beginning work)

____ Plans have been reviewed, including finished basement.

____ Purchase order has been received or picked up.

____ Rough inspections are completed and the house has passed inspections.

____ House is clean and free of debris.

____ All windows are intact and unbroken.

____ Exterior doors are weatherstripped and sealed.

____ All windows are weatherstripped and sealed.

____ Bottom plate is sill sealed.

____ Temporary stairs, if required, are in place.

____ Temporary handrails are in place.

____ Temporary safety bracing is in place.

____ HVAC ductwork is installed.

____ Ducts are covered and protected.

____ Plumbing pipes, tubs, showers, etc., are installed and undamaged.

____ Electrical wiring and boxes are in place.

____ Exterior insulation and sheathing board are in place, unbroken, undamaged, or repaired with no gaps or missing pieces.

Date: _____   Site Superintendent _____

Trade Contractor _____

Final Inspection: (to be completed before Trade Contractor leaves jobsite)

____ All windows are intact.

____ Tubs, showers, etc., are undamaged.

____ All temporary handrails and safety bracing are intact.

____ All penetrations to the outside are sealed.

____ A minimum of three (3) rulers are clearly visible in the attic and are in contact with the ceiling decking.

____ All wall batts fit snugly at the top, bottom, and sides, and are fastened correctly and cover all areas.

____ Vaults and/or tray ceilings have batt insulation.

____ Insulation is of the correct R-value in all areas.

____ Basement walls are insulated including the block foundation if the basement is finished.

____ Ceilings of garages and/or basements that are under heated areas are insulated.

____ Walls between unheated areas and heated areas are insulated.

____ All attic access areas are weatherstripped.

____ Ceiling plates and headers are sealed.

____ Insulation has been placed behind wiring, not stuffed around or covering wiring.

____ Insulation behind receptacle boxes and switch boxes does not block the front.

____ All windows and exterior doors have insulation in jacks and headers.

____ House is clean and broom-swept.

____ All building debris has been removed to dumpster.

Date: _____   Site Superintendent _____

Trade Contractor _____

<u>SCOPE OF WORK</u>
Standards and Description of Work Performance

LANDSCAPING

The Company's **Terms and Conditions** are by reference a part of all **Scope of Work** requirements.

Construction Requirements:

Generally speaking the work of Trade Contractors and their employees is expected to be performed in a good and workmanlike manner. Workmanlike quality is defined as workmanship that meets or betters those criteria indicated in applicable building codes, using materials and installation methods identified in the construction plans and this Scope of Work.

Code Requirements:

All jobs shall conform to those standards stipulated in the building code, mechanical code, plumbing code, and electrical code applicable in the local jurisdiction. All construction on The Company's jobsites shall meet or exceed NAHB Performance and Building Standards.

General Comments:

The Company considers our Trade Contractors to be experts at producing a high-quality job. But everyone on our construction team—staff, Trade Contractors, and suppliers—must recognize the importance of providing quality in both the product and service areas while on our jobsites and in the homes of our purchasers.

Since we work as a team, poor quality or service, from any of us, reflects unfavorably on all of us. An exceptional level of product quality and highly effective service can help us all to increase our business and grow.

The Company's definition of quality construction also requires that every job be completed correctly the first time. When this does not occur it costs both of us additional money, imposes on the purchaser, and hurts our reputations as quality builders. That is why, in situations where construction was not completed in a quality manner, prompt corrective action is required to remedy specific deficiencies.

In the following information the term Site Superintendent shall refer to any The Company representative with authority to perform the specified task. The term Trade Contractor shall mean the Trade Contractor's organization or any representative that is assigned the authority to perform the specified task.

General: The landscaping Trade Contractor is responsible for the appearance, drainage, and greenery of the house. Depending on the subdivision, The Company's houses receive sod or seed or a combination of sod and seed.

Company Rep's Initials _____
Trade Contractor's Initials _____

Materials: The Trade Contractor's purchase order shall specify whether the house will receive sod, seed, or a combination of sod and seed and what area the sod, if specified, shall cover. The Company shall determine the number and size of trees, shrubs, and/or flowers to be installed.

Installation: All work shall be done by experienced, trained personnel. Prior to any yard receiving seed or sod the yard shall be fine-graded and all rocks, unbreakable clumps of earth, construction debris, or other debris shall be removed from yard. The grading of the yard shall be such that all swales or berms are finalized to ensure the correct flow of water. Soil shall be leveled to remove any gullies, washes, large depressions, and humps.

Sod shall be applied with no gaps between sod squares. Sod shall be rolled upon completion of the installation of sod. Grass seed shall be applied at 50 pounds per 5,000 square feet. Fertilizer shall be of a type and amount required by the location and season of the year. Straw shall be applied in an amount sufficient to protect seed and fertilizer.

Shrubs and trees shall be of a size and number specified by The Company. All shrubs and trees shall be planted in soil that has been loosened prior to planting. Once shrubs and trees are planted the soil should be packed to eliminate all air pockets and then watered.

Warranty: The Company's printed Limited Warranty booklet excludes landscaping from the Warranty coverage except in relation to standing water and improper drainage. However, The Company believes that all work done in connection with one of our homes should be of quality work and all Trade Contractors should stand behind the quality of their work. Therefore, we require all Trade Contractors who furnish landscaping to warrant their work for a period of one (1) year from the closing of the house.

The Trade Contractor shall have seven (7) days in which to correct any Warranty problem. If the problem is not corrected within seven (7) days then The Company shall correct the problem and will backcharge the Trade Contractor at the rate of $25.00 per hour, with a minimum charge of $100.00 plus the cost of any materials.

Inspection Reports: The Trade Contractor and the Site Superintendent shall walk the job together and complete each section of the inspection report(s). The Trade Contractor must correct any deficiency found during the inspection and the job must be 100-percent complete before payment will be made. The Trade Contractor and the Site Superintendent must sign-off on all sections of the inspection report(s) attesting that the job is 100-percent complete and is correct per the job requirements found in this Scope of Work.

Detailed Job Requirements:

1. Prior to work beginning the Trade Contractor shall review the plans with the Site Superintendent to ensure the Trade Contractor understands the requirements for the house or lot.

2. Purchase orders shall be picked up from the Site Superintendent.

3. The Trade Contractor and the Site Superintendent must walk the job together and complete the pre-work section of the inspection report(s) before work may begin. The pre-work section of the inspection report must be signed-off on by both parties.

_____ Company Rep's Initials
_____ Trade Contractor's Initials

4. The Trade Contractor is to remove all formboards from drives, walks, patios, etc. Formboard material shall be stacked in the garage to be reused.

5. The Trade Contractor shall fine-grade the yard area to be landscaped to properly manage the water flow.

6. If any swale or berm has been damaged or is incorrect to ensure correct water flow (6 inches of fall within 10 feet of the house) the Site Superintendent should be notified immediately.

7. The Trade Contractor shall take all precautions to ensure no mechanicals are damaged during the fine-grading of the lot.

8. During the fine-grade the Trade Contractor shall ensure soil is a minimum of 8 inches from the bottom of the first piece of siding on wall.

9. The yard shall be fine-raked to remove any construction debris, gravel, rocks, unbreakable lumps of soil, and any other foreign debris that would interfere with the smooth, pleasing appearance of the yard.

10. If the purchase order calls for sod, no sod shall be installed over washes, gullies, or large lumps. Sod shall be laid with no gaps between squares. Sod shall be sod stapled in such a manner to prevent movement of sod. All planting areas shall be formed in such a manner that the edge of the sod will be smooth and present a uniform appearance. Jagged edges or an out-of-shape appearance is unacceptable. At the completion of the installation all sod shall be rolled. If weather is too dry to roll the sod, water shall be applied and sod shall be rolled before job will be considered complete.

11. Seed shall be applied at 50 pounds per 5,000 square feet of soil.

12. Fertilizer shall be applied as required by the area and the season of the year.

13. Straw shall be applied to completely cover the seeded area.

14. All sod and seed shall be watered immediately after installation.

15. Planter areas shall have the soil loosened to ensure survival for shrubs and trees.

16. Shrubs and trees shall be of the type and number specified on the purchase order.

17. Shrubs and trees shall be planted in loosened soil, and then firmly packed to eliminate air pockets and watered.

18. All planter areas shall be covered with pine straw to protect new plants. Pine straw should enhance the appearance of the planted area not distract from it.

19. Any areas that do not require grass should have pine straw applied to prevent weeds and to enhance the appearance of the area (such as under decks, etc.).

20. The Trade Contractor and Site Superintendent must walk the job together and perform a final inspection of the job. The final section of the inspection report(s) must be completed and

Company Rep's Initials _____
Trade Contractor's Initials _____

signed-off on by both parties. The inspection report(s) must be attached to the office's copy of the purchase order and the Trade Contractor's invoice or payment will not be issued.

21. Any items found during the final inspection that need correction shall be corrected before payment will be made.

I _____ agent for _____

_____ have read and fully understand the above **Scope of Work** and I hereby agree to perform all work in accordance with the above.

Date: _____   _____
                            Signed: Trade Contractor (or agent)

Date: _____   _____
                            For The Company

INSPECTION REPORT

LANDSCAPING

Subdivision/Lot # _____

Pre-Work Inspection: (to be completed prior to beginning work)

____ Plans have been reviewed with Site Superintendent and the lot walked.

____ Purchase order has been received.

____ Yard area is clean and free of debris.

____ Swales and berms are established.

____ All lines, pipes, etc., are buried.

____ Temporary power pole is removed.

____ Drive and walks are undamaged.

Date: _____   Site Superintendent _____

                         Trade Contractor _____

Final Inspection: (to be completed before Trade Contractor leaves jobsite)

____ Formboards are removed and stacked in garage.

____ Drives and walks are undamaged.

____ Swales and berms are finalized.

____ Fall is 6 inches of fall within 10 feet of house.

____ There is a minimum of 8 inches between bottom of siding and final grade in all areas.

____ Water flow and direction are correct.

____ There are no low areas that will permit the pooling of water.

____ Yard has a level, attractive appearance.

____ There are no obvious hills, gullies, washes, low spots, or bulges in yard areas.

____ Sod, if required, is healthy, with no gaps. Sod is stapled and rolled.

____ Yard is seeded. Seed can be seen clearly when straw is moved out of way.

____ Fertilizer has been applied. Fertilizer can be seen when straw is moved out of way.

____ Sod and seed have been watered.

____ Planter areas are neatly edged and pine straw applied.

____ Trees and shrubs are firmly planted, healthy looking, and watered.

____   Areas not requiring grass have had pine straw applied and are neat in appearance.

____   Yard is clean and free of trash and debris.

Date: _____   Site Superintendent _____

                              Trade Contractor _____

SCOPE OF WORK
Standards and Description of Work Performance

MIRRORS

The Company's **Terms and Conditions** are by reference a part of all **Scope of Work** requirements.

Construction Requirements:

Generally speaking the work of Trade Contractors and their employees is expected to be performed in a good and workmanlike manner. Workmanlike quality is defined as workmanship that meets or betters those criteria indicated in applicable building codes, using materials and installation methods identified in the construction plans and this Scope of Work.

Code Requirements:

All jobs shall conform to those standards stipulated in the building code, mechanical code, plumbing code, and electrical code applicable in the local jurisdiction. All construction on The Company's jobsites shall meet or exceed NAHB Performance and Building Standards.

General Comments:

The Company considers our Trade Contractors to be experts at producing a high-quality job. But everyone on our construction team—staff, Trade Contractors, and suppliers—must recognize the importance of providing quality in both the product and service areas while on our jobsites and in the homes of our purchasers.

Since we work as a team, poor quality or service, from any of us, reflects unfavorably on all of us. An exceptional level of product quality and highly effective service can help us all to increase our business and grow.

The Company's definition of quality construction also requires that every job be completed correctly the first time. When this does not occur it costs both of us additional money, imposes on the purchaser, and hurts our reputations as quality builders. That is why, in situations where construction was not completed in a quality manner, prompt corrective action is required to remedy specific deficiencies.

In the following information the term Site Superintendent shall refer to any The Company representative with authority to perform the specified task. The term Trade Contractor shall mean the Trade Contractor's organization or any representative that is assigned the authority to perform the specified task.

General: Mirrors will be of the number, type, and size specified per plan and purchase order.

Installation: Installation shall be done by experienced, trained personnel. All mirrors shall be installed plumb, level, and square, without damage to the mirror or surrounding areas. All mirrors

Company Rep's Initials _____
Trade Contractor's Initials _____

shall be installed per the manufacturer's installation instructions and with the type of fasteners approved by the manufacturer.

Installation shall ensure that there is no damage to drywall or other finished work. Mirrors shall not be installed if they are chipped, scratched, cracked, or if the backing has been damaged.

Warranty: The Company believes that all work done and all materials installed in connection with one of our homes should be of high quality and that all Trade Contractors should stand behind the quality of their work and materials. Therefore we require all Trade Contractors to warrant the quality of their work for a period of one (1) year from date of closing of the house. Please refer to our printed Limited Warranty booklet for any specific items that are covered under Warranty as they apply to mirrors.

The Trade Contractor shall have seven (7) days in which to correct any Warranty problem. If the problem is not corrected within seven (7) days then The Company shall correct the problem and will backcharge the Trade Contractor at the rate of $25.00 per hour, with a minimum charge of $100.00 plus the cost of any materials.

Inspection Reports: The Trade Contractor and the Site Superintendent shall walk the job together and complete each section of the inspection report(s). The Trade Contractor must correct any deficiency found during the inspection and the job must be 100-percent complete before payment will be made. The Trade Contractor and the Site Superintendent must sign-off on all sections of the inspection report(s) attesting that the job is 100-percent complete and is correct per the job requirements found in this Scope of Work.

Detailed Job Requirements:

1. Plans are subject to changes and modifications. It is the responsibility of the Trade Contractor to verify with the Site Superintendent that the plans have not changed before beginning work. The Trade Contractor at no cost to The Company will correct any errors that occur from using an incorrect set of plans.

2. The purchase order shall be mailed from the office. If the purchase order has not been received prior to installation, a copy of the purchase order should be picked up from the Site Superintendent.

3. The Trade Contractor and the Site Superintendent must walk the job together and complete the pre-work section of the inspection report(s) before work may begin. The pre-work section of the inspection report(s) must be signed-off on by both parties.

4. Mirrors shall be the size, number, and type specified by The Company. No substitutions will be allowed.

5. All mirrors shall be installed per the installation instructions of the manufacturer with the suggested fasteners.

6. All mirrors shall be installed per plan.

7. All mirrors shall be installed level, plumb, and square with no damage to drywall or existing work.

_____ Company Rep's Initials
_____ Trade Contractor's Initials

8. House shall be left clean, broom-swept, and all debris shall be removed to the dumpster or to the designated trash site.

9. The Trade Contractor and Site Superintendent must walk the job together and perform a final inspection of the job. The final section of the inspection report(s) must be completed and signed-off on by both parties. The inspection report(s) must be attached to the office's copy of the purchase order and the Trade Contractor's invoice or payment will not be issued.

10. Any items found during the final inspection that need correction, shall be corrected before payment will be made.

I _____ agent for _____

_____ have read and fully understand the above **Scope of Work** and I hereby agree to perform all work in accordance with the above.

Date: _____

_____
Signed: Trade Contractor (or agent)

Date: _____

_____
For The Company

INSPECTION REPORT

MIRRORS

Subdivision/Lot # _____

Pre-Work Inspection: (to be completed prior to beginning work)

____   Purchase order has been received.

____   Plans have been verified.

____   Drywall and painting are complete and ready for mirrors.

____   Cabinets are installed.

____   Drywall and paint are undamaged.

____   Cabinet tops are installed, undamaged, and protected.

____   Windows are installed and undamaged.

____   House is clean, broom-swept, and free of debris.

Date: _____   Site Superintendent _____

                                         Trade Contractor _____

Final Inspection: (to be completed before Trade Contractor leaves jobsite)

____   All windows are intact and unbroken.

____   Drywall and paint are undamaged.

____   Countertops are undamaged and protected.

____   Mirrors are installed level, plumb, and square.

____   Mirrors are installed per plan.

____   Mirrors are undamaged with no chips, cracks, scratches, or damage to backing.

____   House is clean and broom-swept.

____   Debris has been removed to correct area.

Date: _____   Site Superintendent _____

                                         Trade Contractor _____

SCOPE OF WORK
Standards and Description of Work Performance

PAINTING

The Company's **Terms and Conditions** are by reference a part of all **Scope of Work** requirements.

Construction Requirements:

Generally speaking the work of Trade Contractors and their employees is expected to be performed in a good and workmanlike manner. Workmanlike quality is defined as workmanship that meets or betters those criteria indicated in applicable building codes, using materials and installation methods identified in the construction plans and this Scope of Work.

Code Requirements:

All jobs shall conform to those standards stipulated in the building code, mechanical code, plumbing code, and electrical code applicable in the local jurisdiction. All construction on The Company's jobsites shall meet or exceed NAHB Performance and Building Standards.

General Comments:

The Company considers our Trade Contractors to be experts at producing a high-quality job. But everyone on our construction team—staff, Trade Contractors, and suppliers—must recognize the importance of providing quality in both the product and service areas while on our jobsites and in the homes of our purchasers.

Since we work as a team, poor quality or service, from any of us, reflects unfavorably on all of us. An exceptional level of product quality and highly effective service can help us all to increase our business and grow.

The Company's definition of quality construction also requires that every job be completed correctly the first time. When this does not occur it costs both of us additional money, imposes on the purchaser, and hurts our reputations as quality builders. That is why, in situations where construction was not completed in a quality manner, prompt corrective action is required to remedy specific deficiencies.

In the following information the term Site Superintendent shall refer to any The Company representative with authority to perform the specified task. The term Trade Contractor shall mean the Trade Contractor's organization or any representative that is assigned the authority to perform the specified task.

General: For most buyers the look of the finished paint defines quality construction. The look and feel of the walls, trim, and exterior is a major selling point of most homes. The first impression of the home is created by the look of the exterior and carriers over into the interior with the feel of a clean, fresh, sparkling house. Workmanship that is not high quality turns off the homebuyer.

Company Rep's Initials _____
Trade Contractor's Initials _____

Painted or finished surfaces should present a smooth, unblemished, homogeneous appearance without drops, runs, lumps, streaking, or visible color variations. Each coat should be allowed to dry prior to the application of the next coat. Exterior paints or stains should not peel or deteriorate during the first year of Warranty coverage. Natural finishes on interior woodwork should not deteriorate during the first year of Warranty coverage.

Materials: The Company will determine the manufacturer and grade of paint to be used in all houses.

Paints and solvents should be maintained at a temperature between 50 degrees Fahrenheit (F) and 90 degrees F, and stored in a well-ventilated area. Freezing temperatures may permanently damage water-based paints as may subjecting paint to frost.

Paints not stored within the recommended temperature range should be conditioned for at least 24 hours at a temperature of 65 degrees F to 85 degrees F. Paint materials should be mixed prior to delivery to the job and then hand-mixed just prior to use and periodically during application. Caution should be taken not to over mix, causing the incorporation of excess air.

Preparation: Surfaces intended for coating should be clean, sound, and uniform in nature. To achieve maximum coating life, surfaces should be cleaned of dirt, grease, rust, and moisture. Sharp edges, irregular areas, cracks, and holes should be repaired before application.

When filling masonry, plaster, wood, or wallboard, the area should be cleared of loose debris. Apply compound with a putty knife or trowel and smooth off the surface so it is slightly convex to allow for shrinkage. Damaged areas outside the scope of painting work should be brought to the attention of the Site Superintendent immediately.

All nicks, gouges, scrapes, and damage to trim should be filled with wood filler, sanded, and smoothed. All nail holes shall be filled with wood filler, sanded, and smoothed. Any nails that are protruding shall be set, then filled, sanded, and smoothed.

All work is to be done by trained, experienced individuals. Surface and air temperatures should be between 50 and 90 degrees F for water-based coatings and 45 to 95 degrees F for other coatings unless the manufacturer stipulates otherwise.

Paint should be maintained at 65 to 85 degrees F at all times during application. Paint should not be applied when temperature is expected to freeze prior to drying. Paints should be applied at manufacturer's spreading rates. When successive coats are used, allow sufficient time for each coat to dry thoroughly before the following coat is applied. Materials below or adjoining the work should be covered or otherwise protected.

Payment in stages: The Trade Contractor shall be paid in three stages: rough, final, and touch-up. An inspection report must be completed for each section before payment will be approved. Any items found during the final inspection that need correction shall be corrected before payment will be made. The Trade Contractor is expected to perform only one touch-up and that is after the work is completed on the homeowner's walk-through list. Any other touch-up trips require a purchase order issued by The Company's purchasing agent and the Trade Contractor shall be paid for the trip.

_____ Company Rep's Initials
_____ Trade Contractor's Initials

Other: It is generally assumed in the construction industry that the painting Trade Contractor will somehow make a subquality house look like a quality house. We all know that the final appearance of the house starts at the foundation and ends with the painter. It is not the responsibility of the painting Trade Contractor to correct another Trade Contractor's work.

In the Detailed Job Requirements section below, mention is made of correcting scrapes, nicks, etc. It is not expected that the painting Trade Contractor should repair the trim, drywall, or other Trade Contractors' work that can be spotted during the completion of the initial inspection report, but the painting Trade Contractor looks at every square inch of drywall and trim during the course of the job. Small nicks, gouges, a few nails not properly seated, etc., are items that are only noticed during painting and these are the items that the painting Trade Contractor is expected to correct.

The final interior paint job will be inspected in both sunlight and normal room lighting. Any defect visible from a distance of 6 feet under these conditions will be considered unacceptable.

Warranty: The Company believes that all work done and all materials installed, in connection with one of our homes should be of high quality and that all Trade Contractors should stand behind the quality of their work and materials. Therefore we require all Trade Contractors to warrant the quality of their work for a period of one (1) year from date of closing of the house. Please refer to our printed Limited Warranty booklet for specific items that are covered under Warranty as they apply to painting.

The Trade Contractor shall have seven (7) days in which to correct any Warranty problem. If the problem is not corrected within seven (7) days then The Company shall correct the problem and will backcharge the Trade Contractor at the rate of $25.00 per hour, with a minimum charge of $100.00 plus the cost of any materials.

Inspection Reports: The Trade Contractor and the Site Superintendent shall walk the job together and complete each section of the inspection report(s). The Trade Contractor must correct any deficiency found during the inspection and the job must be 100-percent complete before payment will be made. The Trade Contractor and the Site Superintendent must sign-off on all sections of the inspection report(s) attesting that the job is 100-percent complete and is correct per the job requirements found in this Scope of Work.

Detailed Job Requirements:

1. Plans and/or requirements of the job shall be reviewed with the Site Superintendent prior to beginning work.

2. Purchase orders should be picked up from the Site Superintendent.

3. Color selections shall be given to the Trade Contractor in ample time to secure the correct paint. Under no circumstances shall the Trade Contractor change the type or grade of paint specified by The Company.

4. The Trade Contractor and the Site Superintendent must walk the job together and complete the pre-work section of the inspection report(s) before work may begin. The pre-work section of the inspection report(s) must be signed-off on by both parties.

Company Rep's Initials _____
Trade Contractor's Initials _____

5. All nicks, gouges, scrapes, damage, etc., should be repaired, treated, or otherwise taken care of before painting begins, both in the drywall and trim.

6. All nail holes in trim are to be filled with wood filler, sanded, and smoothed. All nail holes in drywall shall be repaired with drywall mud, not caulk.

7. Excess damage to drywall, trim, or doors is to be reported immediately to the Site Superintendent.

8. Interior wall paint shall be a water-based latex paint as specified by The Company.

9. The first coat of interior wall paint can be sprayed as long as surrounding items are protected from paint damage.

10. The second coat of interior wall paint must be rolled or sprayed and then back-rolled.

11. All interior trim paint shall be water-based as specified by The Company.

12. All interior trim paint shall be applied with a brush and shall have two (2) coats.

13. Paint drippings and spills must be cleaned from tubs, showers, countertops, and all other areas.

14. All over-spray on window glass and aluminum window trim (inside and outside) must be wiped off with a damp rag before it dries. Aluminum window trim, especially the white baked-on finish, is easily damaged by razor blades and other items when the cleaners try to remove dried paint over-spray.

15. Paint drippings or spills on vinyl or wood flooring must be cleaned **immediately** without damage to the vinyl or wood flooring.

16. The final interior paint job will be inspected in both sunlight and normal room lighting. Any defect visible from a distance of 6 feet under these conditions will be considered unacceptable.

17. The Trade Contractor is to take care not to damage work done by other Trade Contractors. If any sink is used for washing brushes, etc., the sink shall be cleaned of any paint residue prior to the job being considered complete.

18. Exterior paint shall be as specified by The Company.

19. Exterior paint shall be two (2) coats, both of which may be sprayed.

20. Exterior trim paint shall be as specified by The Company.

21. Exterior trim paint shall be two (2) coats, both of which must be brushed.

22. All exterior areas to be painted, including the foundation, shall be free of dirt, mud, excess caulking, and any other foreign matter.

23. No thin spots in either interior or exterior paint are allowed.

_____ Company Rep's Initials
_____ Trade Contractor's Initials

24. Finishes shall be uniform and smooth, with no lumps, drips, runs, streaking, or visible color variations.

25. Any over-painting on trim must be cleaned from surrounding area.

26. Paint must be cleaned from all hinges and other hardware.

27. All door painting must be smooth, with no lumps, bumps, runs, color variations, and of a uniform appearance. Any defect visible from a distance of 6 feet in sunlight and normal lighting is unacceptable.

28. Over-spray must be cleaned from any brick, stucco, or other finish.

29. Decks, porches, outside handrails, etc., are a part of the outside paint package and shall be painted, stained, or left untreated per plan or purchase order.

30. Any defect in outside paint (walls, trim, etc.) visible from a distance of 6 feet in normal sunlight will be considered unacceptable.

31. Interior handrail coatings shall be uniform with no drips, runs, light spots, etc. Any defect visible from a distance of 6 feet in sunlight and normal lighting is unacceptable.

32. All interior base, casing, shoe molding, door, and window jams, crown molding, chair rail and any other wood trim will be caulked.

33. All exterior siding joints, corner boards, facia boards, overhangs, cantilevers, windows, doors, vents, and any other wood trim or siding shall be caulked. On houses with brick fronts, all gaps between the brickmold and brick will be caulked.

34. All thresholds must be caulked to prevent moisture and insects from entering the house.

35. Stairs that have oak tread areas shall be stained per the color selection sheet. Stains are to be uniform with no light spots and no damage to the wood. All treads should be as close in color as possible to each other.

36. Care should be taken that the stain of the stair treads does not get on pickets nor paint from the pickets on the oak tread area. **Any such overlap must be cleaned immediately while paint or stain is wet.** Any damage that results from cleaning after the paint or stain is dry is the responsibility of the Trade Contractor. If pickets or treads must be replaced because of such damage it will be at cost of the Trade Contractor.

37. All exposed metal lintels shall be wire-brushed to remove any rust and painted with a black, rust-proof paint.

38. All doors that have been taken down to paint shall be rehung, straight, level, and undamaged.

39. All debris shall be removed to the dumpster or to the designated trash area.

40. House shall be clean and broom-swept before job will be considered complete.

Company Rep's Initials _____
Trade Contractor's Initials _____

41. The Trade Contractor and Site Superintendent must walk the job together and perform a final inspection of the job. The final section of the inspection report(s) must be completed and signed-off on by both parties. The inspection report(s) must be attached to the office's copy of the purchase order and the Trade Contractor's invoice or payment will not be issued.

42. Any items found during the final inspection that need correction shall be corrected before payment will be made.

I _____ agent for _____

_____ have read and fully understand the above **Scope of Work** and I hereby agree to perform all work in accordance with the above.

Date: _____

_____
Signed: Trade Contractor (or agent)

Date: _____

_____
For The Company

INSPECTION REPORT

PAINTING (ROUGH)

Subdivision/Lot # _____

Pre-Work Inspection: (to be completed prior to beginning work)

____   Plans have been reviewed with Site Superintendent.

____   Purchase order has been received.

____   Windows are intact with none broken.

____   Vinyl and wood floors are protected.

____   Tubs, showers, countertops, etc., are protected.

____   All drywall is installed, taped, mudded, and sanded.

____   Drywall is straight with no bows or depressions.

____   Joints in drywall are smooth and clean.

____   Drywall has no excessive nicks, gouges, scrapes, etc.

____   Drywall has no raised facepaper from over-sanding.

____   Corners have no hairline cracks.

____   Drywall has no nail pops or loose nails.

____   Trim fits properly around all windows.

____   Trim fits properly around all doors.

____   Base has no nicks, gouges, scratches, damage, etc.

____   Nails are set properly in base at correct depth with no protruding nails.

____   Base is secured tightly, corners are correct and tight.

____   Base around cabinets is tacked in place.

____   All doors are installed according to plan, and are plumb, square, the proper distance for carpet or vinyl, and swing correctly.

____   All bifold doors are square, plumb, and tracks are installed tightly.

____   Attic accesses are installed correctly and all trim is in place.

____   There is no missing trim in any room.

____   Chair rail or crown is installed, if required by plan or options.

____   House is broom-swept, clean, and free of debris.

Date: _____   Site Superintendent _____

Trade Contractor _____

Final Inspection: (to be completed before Trade Contractor leaves jobsite)

____ Exterior wall paint has no light spots and uniform coverage.

____ Exterior trim paint has no paint on walls, uniform coverage, and uniform color.

____ Exterior paint job inspected from 6-foot distance (entire job) with no visible defects.

____ No paint is on windows, brick, stucco, or other areas.

____ Metal lintels are clean (no rust) and painted black with rust-proof paint.

____ Interior walls are smooth with no drips, runs, lumps, bumps, color variations, or streaking.

____ Interior walls have uniform coverage after first coat is sprayed.

____ All rough spots have been repaired in interior trim. Nail holes are filled and sanded smooth.

____ First coat trim paint applied.

____ Excess trim paint on walls has been cleaned off.

____ Over-spray has been cleaned from windows (glass and aluminum).

____ There is no damage to floors from spills, drips, etc.

____ There is no damage to vinyl.

____ All interior wood is caulked.

____ All exterior wood is caulked.

____ All gaps between the brickmold and brick are caulked.

____ All thresholds are caulked.

____ Windows are intact with none broken.

____ Tubs, sinks, showers, etc., have no damage.

____ House is broom-swept, clean, and free of debris.

Date: _____   Site Superintendent _____

                               Trade Contractor _____

INSPECTION REPORT

PAINTING (FINAL)

Subdivision/Lot # _____

Pre-Work Inspection: (to be completed prior to beginning work)

____ There is no damage to tubs, showers, countertops, stairs, windows, vinyl, etc.

____ All protective coverings are in place (tubs, countertops, vinyl, etc.).

____ All drywall repairs have been completed, surface is smooth and sanded with no raised facepaper from over-sanding.

____ All damaged trim has been repaired or replaced.

____ There is no carpet on the floor.

____ House is ready to be painted.

____ House is clean, broom-swept, and free of debris.

Date: _____   Site Superintendent _____

Trade Contractor _____

Final Inspection: (to be completed before Trade Contractor leaves jobsite)

____ Second coat of paint applied to walls is smooth, with no runs, drips, lumps, color variations, streaking, or light spots.

____ Second coat of trim paint applied to trim and doors is smooth, with no runs, drips, lumps, color variations, streaking, or light spots.

____ All stairs and handrails are stained or painted per plan.

____ All stained areas have a uniform appearance and complete coverage.

____ Varnish is applied smoothly and uniformly to all stained areas.

____ There is no damage to drywall, stairs, handrails, trim, etc.

____ Paint spills have been **removed without damage** to vinyl, wood floors, tubs, showers, countertops, etc.

____ Paint has been removed from window glass and aluminum trim.

____ Paint has been removed from door hinges and all hardware.

____ Doors have been rehung, and are square, level, plumb, and back where they belong.

____ Thresholds are stained at all exterior doors.

____ No defects are visible under sunlight and normal house lighting from a distance of 6 feet for the entire interior paint job.

____   There is no paint residue in sinks.

____   Debris has been removed to the dumpster.

____   House is clean and broom-swept.

____   Excess material is stored in garage.

Date: _____   Site Superintendent _____

                                         Trade Contractor _____

INSPECTION REPORT

PAINTING (TOUCH-UP)

Subdivision/Lot # _____

Pre-Work Inspection: (to be completed prior to beginning work)

____   Trade Contractor and Site Superintendent have walked the house to determine touch-up areas.

____   All areas to be touched up are plainly marked with colored dots or the Trade Contractor has received a punch-list.

____   All work by other trade contractors is complete, except the cleaning crew.

Date: _____   Site Superintendent _____

                           Trade Contractor _____

Final Inspection: (to be completed before Trade Contractor leaves jobsite)

____   All areas of paint have been inspected and no missed repairs were found.

____   Touch-up paint areas blend with surrounding areas (no difference visible from 6 feet in sunlight from windows and under normal house lighting).

____   No excess paint is on any area (windows, trim, doors, floors, etc.).

____   House is clean and free of debris, and ready for the cleaning crew.

Date: _____   Site Superintendent _____

                           Trade Contractor _____

SCOPE OF WORK
Standards and Description of Work Performance

PLUMBING

The Company's **Terms and Conditions** are by reference a part of all **Scope of Work** requirements.

Construction Requirements:

Generally speaking the work of Trade Contractors and their employees is expected to be performed in a good and workmanlike manner. Workmanlike quality is defined as workmanship that meets or betters those criteria indicated in applicable building codes, using materials and installation methods identified in the construction plans and this Scope of Work.

Code Requirements:

All jobs shall conform to those standards stipulated in the building code, mechanical code, plumbing code, and electrical code applicable in the local jurisdiction. All construction on The Company's jobsites shall meet or exceed NAHB Performance and Building Standards.

General Comments:

The Company considers our Trade Contractors to be experts at producing a high-quality job. But everyone on our construction team—staff, Trade Contractors, and suppliers—must recognize the importance of providing quality in both the product and service areas while on our jobsites and in the homes of our purchasers.

Since we work as a team, poor quality or service, from any of us, reflects unfavorably on all of us. An exceptional level of product quality and highly effective service can help us all to increase our business and grow.

The Company's definition of quality construction also requires that every job be completed correctly the first time. When this does not occur it costs both of us additional money, imposes on the purchaser, and hurts our reputations as quality builders. That is why, in situations where construction was not completed in a quality manner, prompt corrective action is required to remedy specific deficiencies.

In the following information the term Site Superintendent shall refer to any The Company representative with authority to perform the specified task. The term Trade Contractor shall mean the Trade Contractor's organization or any representative that is assigned the authority to perform the specified task.

General: The plumbing system shall be a complete and functioning system tested and ready for operation.

The system and fixtures shall be installed in accordance with all approved building, plumbing, and health codes, and function as specified. Installations should pass the initial compliance inspection at each phase.

_____ Company Rep's Initials
_____ Trade Contractor's Initials

Service connections to water main and sewer should function properly. Piping shall be designed to ensure that there is adequate pressure and flow at each fixture and that each fixture is protected from freezing.

No valve, faucet, or fixture shall leak due to defects in materials or workmanship, nor should leaks of any kind exist in any soil, waste, vent, or water pipe. Noise from loose pipes or a water hammer is unacceptable.

The plumbing system should operate correctly for the first two (2) years of Warranty coverage with the exception of those items caused by owner negligence.

Materials: Fixtures, appliances, and fittings shall be as specified and comply with their manufacturers' standards for performance and installation. The surface of bathtubs and kitchen sinks shall not be chipped or scratched.

The Company's Request for Quote bid sheet shall specify the manufacturer of tubs, sinks, and faucets. Many times these items are dependent on the subdivision in which they are installed. If The Company changes its requirements, the Trade Contractor will be notified in ample time to make arrangements to change manufacturer(s).

Installation: All work is to be done by trained, experienced individuals. Workmanship shall be so as to require a minimum of repairs and patching after installation. Notching, drilling, and cutting of framing members should be done so as not to compromise the structural integrity of the house. **Under no circumstances shall any roof or floor truss be cut, notched, or otherwise damaged**. Should any truss not accommodate the necessary plumbing pipes, fixtures, etc., the Site Superintendent should be notified immediately. The Site Superintendent also must be notified if any truss is accidentally damaged. Repairs cannot be made without the approval of the manufacturer's engineer.

Drain, waste, vent, and water pipes shall be installed in a manner that allows room for adequate insulation as required by applicable code(s) during normally anticipated cold weather, and as defined in accordance with ASHRAE design temperatures, to prevent freezing.

Extreme caution should be exercised while using open flames in buildings.

The Trade Contractor is responsible for final inspections to ensure a quality job.

Warranty: Certain items are warrantable for one (1) year and some for two (2) years. Refer to The Company's printed Limited Warranty booklet for a list of these items. The Trade Contractor shall furnish to the homeowner, via a tag on the hot water tank and on the sticker on the inside of the left door of the cabinet located to the left the kitchen sink, a regular office hour phone number and an emergency phone number. The Trade Contractor is required to furnish emergency service.

The Trade Contractor shall have seven (7) days in which to correct any Warranty problem. If the problem is not corrected within seven (7) days then The Company shall correct the problem and will backcharge the Trade Contractor at the rate of $25.00 per hour, with a minimum charge of $100.00 plus the cost of any materials.

Company Rep's Initials _____
Trade Contractor's Initials _____

Inspection Reports: The Trade Contractor and the Site Superintendent shall walk the job together and complete each section of the inspection report(s). The Trade Contractor must correct any deficiency found during the inspection and the job must be 100-percent complete before payment will be made. The Trade Contractor and the Site Superintendent must sign-off on all sections of the inspection report(s) attesting that the job is 100-percent complete and is correct per the job requirements found in this Scope of Work.

Detailed Job Procedures:

1. A new set of plans is required for each house. Plans are subject to changes and modifications. It is the responsibility of the Trade Contractor to have the new plans before beginning work. Plans should be picked up at the job trailer from the Site Superintendent. The Trade Contractor at no cost to The Company will correct any errors that occur from using an incorrect set of plans.

2. The purchase order for the job will be mailed to the Trade Contractor. If the purchase order is not received a copy should be obtained from the Site Superintendent.

3. The Trade Contractor must be licensed in the county in which the subdivision is located before work may begin.

4. It is the responsibility of the Trade Contractor to secure all necessary water and sewer permits.

5. The Company will specify the manufacturer or brand for all fixtures for the house. This manufacturer or brand may not be changed without approval of The Company. The change must be approved in writing and the Trade Contractor must furnish The Company with copies of specification sheets and warranty material to be included in The Company's New Homeowner's Manual for all fixtures placed in the house.

6. The Trade Contractor and the Site Superintendent must walk the job together and complete the pre-work section of the inspection report(s) before work may begin. The pre-work section of the inspection report(s) must be signed-off on by both parties.

7. All garden tubs, shower stalls, and tub/shower combination units are to placed inside the bathroom areas by the framing Trade Contractor prior to framing the bath areas. The plumbing Trade Contractor shall work with the Site Superintendent to ensure that the fixtures are onsite in time for the framing Trade Contractor to set them inside the house. The plumbing Trade Contractor should not need to adjust any framing members. Should adjustments be necessary the Trade Contractor should notify the Site Superintendent before adjustments are made.

8. Plumbing is divided into three sections: slab, rough, and final. Payment will be made according to these three divisions.

9. The Trade Contractor shall dig all water and sewer tap lines and lay all water and sewer piping. All under-slab water-distribution piping shall be copper water tubing Type M.

10. All water-distribution piping in the house shall be copper water tubing, and all discharge piping shall be polyvinyl chloride (PVC) plastic pipe.

11. All plumbing fixtures shall have approved strainers and/or stoppers, other than water closets.

_____ Company Rep's Initials
_____ Trade Contractor's Initials

12. Fixtures having concealed tubular traps shall be provided with an access panel or unobstructed utility space of at least 12 inches in dimension. Joints that are soldered, screwed, fused, or solvent-welded must form a solid connection.

13. Floor-outlet or floor-mounted fixtures shall be secured to the drainage connection and to the floor, when so designed, by screws, bolts, washers, nuts, and similar fasteners of copper, brass, or other corrosion-resistant materials.

14. Where fixtures come in contact with walls or floors the join shall be watertight.

15. Plumbing fixtures shall be functionally accessible.

16. The water closet centerline shall be not less than 15 inches from adjacent walls, partitions, or cabinets.

17. The location of piping, fixtures, or equipment shall not interfere with window and door operations.

18. Plumbing fixtures or other receptors receiving the discharge of indirect waste pipes shall be shaped and have a capacity to prevent splashing or flooding, and shall be readily accessible for inspection and cleaning.

19. In any house with a basement area, either finished or unfinished, the Trade Contractor shall provide a box for a sewer pump if the floor is below the sewer line.

20. If the plans call for a bath to be stubbed in the basement the Trade Contractor shall cover those items in the slab (drains, pipe, etc.) with a bucket to protect the items. The bucket should be flagged in a manner that makes it readily visible to other Trade Contractors.

21. All fixtures, faucets, showerheads, etc., shall be of the water-conserving type.

22. All walls surrounding a shower compartment or tub/shower combination shall be sealed to form a watertight join with such fixtures.

23. All shower compartments and tub/shower combination fixtures shall be lined per code.

24. All faucets and showerheads shall be correctly aligned so that each faucet/showerhead is the same distance from the wall, and shall be level and straight.

25. All plumbing fixtures (water closet, showers, shower tub units, sinks, etc) shall conform to and be installed to code.

26. All bathtubs shall have outlets and overflows at least 1-1/2 inch in diameter and the waste outlet shall be equipped with an approved stopper.

27. Sinks shall be provided with waste outlets not less than 1-1/2 inch in diameter. A strainer shall be provided to restrict the clear opening of the waste outlet. The sink strainer and rubber stopper for the garbage disposal opening shall be placed in a drawer in the kitchen cabinets to protect them from damage. They are **not** to be left in the sink(s).

28. If a whirlpool tub is to be installed a door or panel of sufficient size shall be installed to provide access to the pump for repair and/or replacement.

Company Rep's Initials _____
Trade Contractor's Initials _____

29. The flow of hot water shall be to the left-hand side of the fittings.

30. Each house shall have an accessible main shutoff valve near the entrance of the water service. The valve shall be of a full-way type with provision for drainage, such as a bleed orifice or installation of a separate drain valve. Additionally, the water service shall have a valve at the curb or property line in accordance with local water authority requirements.

31. A label stating the Trade Contractor's office hour phone number, address, and emergency phone number shall be affixed to the hot water tank and to the left door of the cabinet located to the left of the kitchen sink.

32. The hot water tank shall be a 40-gallon tank unless an approved change order is issued to upgrade the size of the tank.

33. The hot water tank shall be raised 18 inches from the floor. Each plumber must furnish a metal stand manufactured specifically to support a hot water tank. The stand should be installed per the manufacturer's instructions. No tank shall rest flat on a concrete floor or wooden stand.

34. All trash and building debris shall be removed to the dumpster or designated trash area.

35. The house must pass all plumbing inspections. Any corrections required by the inspectors will be at the expense of the Trade Contractor.

36. The Trade Contractor is responsible for cleaning up of all residual materials before the job will be accepted as complete. The house shall be broom-swept before job shall be considered complete.

37. The Trade Contractor and Site Superintendent must walk the job together and perform a final inspection of the job. The final section of the inspection report(s) must be completed and signed-off on by both parties. The inspection report(s) must be attached to the office's copy of the purchase order and the Trade Contractor's invoice or payment will not be issued.

38. Any items found during the final inspection that need correction shall be corrected before payment will be made.

I _____ agent for _____

_____ have read and fully understand the above **Scope of Work** and I hereby agree to perform all work in accordance with the above.

Date: _____   _____
                          Signed: Trade Contractor (or agent)

Date: _____   _____
                          For The Company

INSPECTION REPORT

PLUMBING (SLAB)

Subdivision/Lot # _____

Pre-Work Inspection: (to be completed prior to beginning work)

____ Plans have been picked up and signed for.

____ Purchase order has been received or picked up.

____ Work area is clean and free of debris, ready to dig lines.

____ Foundation is complete, and pad area is ready.

____ Electricity is at site.

Date: _____   Site Superintendent _____

Trade Contractor _____

Final Inspection: (to be completed before Trade Contractor leaves jobsite)

____ Plumbing is set in slab, and capped and protected as required.

Date: _____   Site Superintendent _____

Trade Contractor _____

INSPECTION REPORT

PLUMBING (ROUGH)

Subdivision/Lot # _____

Pre-Work Inspection: (to be completed prior to beginning work)

____ Purchase order has been received.

____ Framing items for plumbing are complete.

____ Temporary stairs are in place, if required.

____ Temporary handrails are in place.

____ Safety bracing is in place.

____ Tubs are placed in bathroom areas.

____ Deadwood has been installed behind shower stall for door installation.

____ Blocking or deadwood has been installed as required.

____ House is clean, broom-swept, and free of prior Trade Contractor's debris.

Date: _____   Site Superintendent _____

                         Trade Contractor _____

Final Inspection: (to be completed before Trade Contractor leaves jobsite)

____ Lines are dug and completed.

____ Tubs, showers, etc., are set.

____ All piping is installed.

____ Drains are installed.

____ No solder or flux is on floor areas.

____ All safety bracing, temporary handrails, etc., are still in place.

____ Rough plumbing inspection has been passed.

____ House is clean and free of debris.

____ House is broom-swept.

Date: _____   Site Superintendent _____

                         Trade Contractor _____

INSPECTION REPORT

PLUMBING (FINAL)

Subdivision/Lot # _____

Pre-Work Inspection: (to be completed prior to beginning work)

____ House is clean and free of debris from prior trade contractors.

____ Windows are intact and unbroken.

____ Vinyl is installed and undamaged.

____ Tub(s) is undamaged.

____ Cabinets are installed.

____ Correct dimensions for water closets (walls, cabinets, etc.) are verified.

____ All safety bracing, temporary handrails, etc., are in place, if required.

Date: _____   Site Superintendent _____

                                          Trade Contractor _____

Final Inspection: (to be completed before Trade Contractor leaves jobsite)

____ All fixtures are set.

____ All fixtures and faucets (brand and type) correspond to purchase order and to plans.

____ All fixtures have been tested for operation.

____ No leaks have been found.

____ No excessive noise is heard from water pipes.

____ No surface defects in faucets, tubs or showers have been found.

____ Tub(s) and shower(s) are fully secured to wall.

____ If pedestal sink, water shutoff valves are accessible and usable.

____ All faucets and showerheads are in alignment (level and straight).

____ Regular water shutoffs are all accessible and in working order.

____ Correct number of water shutoffs have been verified.

____ Water shutoff at street works correctly.

____ Hot and cold water shutoffs are tagged.

____ Water shutoff at main is tagged.

____ Clothes washer faucets are working and tagged with hot and cold tags.

____ Correct number of outside faucets have been verified and all are working.

____ Strainer and stopper are in good condition and placed in drawer in kitchen.

____ Emergency key for garbage disposal is in drawer in kitchen with strainer and stopper.

____ Warranty and instruction manuals for hot water tank and garbage disposal are in kitchen drawer to left of range.

____ Hot water tank is working.

____ Hot water tank is on metal hot water tank stand and secured per manufacturer's instructions.

____ Name and phone number tag, including emergency phone number, is affixed to hot water tank and cabinet door.

____ All inspections have been passed.

____ House is clean, broom-swept, and free of debris.

____ Jobsite is clear of debris, boxes, etc.

Date: _____   Site Superintendent _____

                          Trade Contractor _____

<u>SCOPE OF WORK</u>
Standards and Description of Work Performance

POURED WALL FOUNDATIONS

The Company's **Terms and Conditions** are by reference a part of all **Scope of Work** requirements.

Construction Requirements:

Generally speaking the work of Trade Contractors and their employees is expected to be performed in a good and workmanlike manner. Workmanlike quality is defined as workmanship that meets or betters those criteria indicated in applicable building codes, using materials and installation methods identified in the construction plans and this Scope of Work.

Code Requirements:

All jobs shall conform to those standards stipulated in the building code, mechanical code, plumbing code, and electrical code applicable in the local jurisdiction. All construction on The Company's jobsites shall meet or exceed NAHB Performance and Building Standards.

General Comments:

The Company considers our Trade Contractors to be experts at producing a high-quality job. But everyone on our construction team—staff, Trade Contractors, and suppliers—must recognize the importance of providing quality in both the product and service areas while on our jobsites and in the homes of our purchasers.

Since we work as a team, poor quality or service, from any of us, reflects unfavorably on all of us. An exceptional level of product quality and highly effective service can help us all to increase our business and grow.

The Company's definition of quality construction also requires that every job be completed correctly the first time. When this does not occur it costs both of us additional money, imposes on the purchaser, and hurts our reputations as quality builders. That is why, in situations where construction was not completed in a quality manner, prompt corrective action is required to remedy specific deficiencies.

In the following information the term Site Superintendent shall refer to any The Company representative with authority to perform the specified task. The term Trade Contractor shall mean the Trade Contractor's organization or any representative that is assigned the authority to perform the specified task.

General—Concrete: Concrete is subject to various phenomena including applied chemicals and natural elements that can deteriorate the surface. In some cases, however, surface defects such as flaking, scaling, or spallling may be caused by improper finishing. Excessive powdering or chalking may occur due to improper troweling, excess water, or when uncured concrete freezes. It is important that proper precautions and correct techniques be utilized when handling and finishing of concrete in the field to ensure a quality job.

Company Rep's Initials _____
Trade Contractor's Initials _____

Concrete poured walls shall maintain structural integrity and not crack in excess of 1/8 inch in width or displacement. Foundation variance shall not exceed 1/2 inch out of level in 20 feet, with no ridges or depressions in excess of 1/4 inch within any 32-inch measurement. Foundation walls should not be more than 1 inch out of level over the entire surface and not vary more than 1/2 inch out of square when measured along the diagonal of a 6x8x10-foot triangle at any corner.

Concrete veneer should be laid within 3/8 inch of plumb, level, and true to lines in 4 feet. Overall plumb, level, and wall trueness variance shall not exceed 3/4 inch in 20 feet. Concrete walls or veneer shall not crack in excess of 1/4 inch in width or displacement.

Installation: Concrete shall be placed at an appropriate rate so that it can be spread, straightened, and darbied properly. Techniques for handling and placing concrete should ensure that it remains uniform within each batch and from batch to batch. Concrete should not be allowed to run or be worked over long distances, and should not be allowed to drop more than 4 feet.

Holes left by nails or line pins should be pointed while adjacent concrete is green.

#4 rebar reinforced steel shall be placed vertically 6 feet apart and horizontally 3 feet apart throughout the foundation wall area.

Miscellaneous: If temporary power is not yet connected to the lot, the Trade Contractor must furnish a generator. In no instance may the Trade Contractor draw electricity or water from an occupied or completed home. Permission of the Site Superintendent is required for drawing electricity or water from an adjacent jobsite.

Warranty: The Company believes that all work done in connection with one of our homes should be of high quality and that all Trade Contractors should stand behind the quality of their work. Therefore we require all Trade Contractors to warrant the quality of their work for a period of one (1) year from date of closing of the house. Please refer to our printed Limited Warranty booklet for specific items that are covered under Warranty as they apply to concrete and concrete walls.

The Trade Contractor shall have seven (7) days in which to correct any Warranty problem. If the problem is not corrected within seven (7) days then The Company shall correct the problem and will backcharge the Trade Contractor at the rate of $25.00 per hour, with a minimum charge of $100.00 plus the cost of any materials.

Inspection Reports: The Trade Contractor and the Site Superintendent shall walk the job together and complete each section of the inspection report(s). The Trade Contractor must correct any deficiency found during the inspection and the job must be 100-percent complete before payment will be made. The Trade Contractor and the Site Superintendent must sign-off on all sections of the inspection report(s) attesting that the job is 100-percent complete and is correct per the job requirements found in this Scope of Work.

Detailed Job Requirements:

1. A new set of plans is required for each house. Plans are subject to changes and modifications. It is the responsibility of the Trade Contractor to have the new plans before beginning work. Plans

_____ Company Rep's Initials
_____ Trade Contractor's Initials

should be picked up at the job trailer from the Site Superintendent. The Trade Contractor at no cost to The Company will correct any errors that occur from using an incorrect set of plans.

2. Purchase orders should be picked up from the Site Superintendent.

3. The Trade Contractor and the Site Superintendent must walk the job together and complete the pre-work section of the inspection report(s) before work may begin. The pre-work section of the inspection report(s) must be signed-off on by both parties.

4. Foundations shall be built per plan and/or engineer's specifications.

5. Forms shall be set per plan.

6. Poured walls shall be a minimum of 8 inches in width unless otherwise specified in the engineer's specifications.

7. #4 rebar reinforced steel shall be placed vertically 6 feet apart and horizontally 3 feet apart throughout the foundation wall area.

8. The Trade Contractor is responsible for supplying and setting anchor bolts in all poured walls. Bolts shall be 1/2-inch-diameter anchor bolts placed 6 feet on center and not more than 12 inches from corners. Bolts shall extend a minimum of 15 inches into masonry or 7 inches into concrete.

9. All walls shall be true, plumb, and square. Poured walls shall not be out of plumb more than 1-1/2 inches in 8 feet.

10. Poured concrete foundation walls shall maintain structural integrity and not crack in excess of 1/8 inch in width or displacement.

11. Poured wall variance should not exceed 1/2 inch out of level in 20 feet, with no ridges or depressions in excess of 1/4 inch within any 32-inch measurement.

12. Foundation walls shall not be more than 1 inch out of level over the entire surface and not vary more than 1/2 inch out of square when measured along the diagonal of a 6x8x10-foot triangle at any corner.

13. Poured concrete foundation walls shall not leak.

14. There shall be no honeycombs.

15. Concrete poured at temperatures at or below 32 degrees Fahrenheit (F) shall contain 2-percent calcium.

16. Any walls poured in conditions that have the potential for frost shall be covered with polyethylene for protection against frost damage.

17. All trash and debris shall be removed from lot to dumpster or the designated trash area.

18. All runoff concrete/Portland cement shall be removed to the driveway cut.

Company Rep's Initials _____
Trade Contractor's Initials _____

19. Any excess material shall be moved to the storage location, counted, and noted on the purchase order.

20. No invoice will be paid until the Site Superintendent and the Trade Contractor, together, have field-measured and verified all items shown on the Trade Contractor's invoice. The Trade Contractor's invoice must be signed by both the Site Superintendent and the Trade Contractor to certify that the measurements as shown on the invoice are correct and accurate. Payment will be based on the field measurements and not on the amount of concrete delivered to the site.

21. The Trade Contractor and Site Superintendent must walk the job together and perform a final inspection of the job. The final section of the inspection report(s) must be completed and signed-off on by both parties. The inspection report(s) must be attached to the office's copy of the purchase order and the Trade Contractor's invoice or payment will not be issued.

22. Any items found during the final inspection that need correction shall be corrected before payment will be made.

I _____ agent for _____

_____ have read and fully understand the above **Scope of Work** and I hereby agree to perform all work in accordance with the above.

Date: _____   _____
                            Signed: Trade Contractor (or agent)

Date: _____   _____
                            For The Company

INSPECTION REPORT

POURED WALL FOUNDATIONS

Subdivision/Lot # _____

Pre-Work Inspection: (to be completed prior to beginning work)

____ Plans have been picked up and signed for.

____ Purchase order has been picked up.

____ Work area is clean and free of debris.

____ Footing(s) is poured, and alignment, square corners, etc., are verified.

____ Dimensions are correct.

____ Stakes are set and locations are correct.

____ Straightness (alignment) of lines has been verified.

____ All setbacks, bay windows, fireplace(s), porches, stoops, etc., are clearly marked and dimensions have been checked.

____ All materials are on site and ready to be used.

____ Area is clear for concrete truck.

Date: _____   Site Superintendent _____

                           Trade Contractor _____

Final Inspection: (to be completed before Trade Contractor leaves jobsite)

____ All walls are true, plumb, and square.

____ There are no cracks.

____ Foundation variance does not exceed 1/2 inch out of level in 20 feet, with no ridges or depressions in excess of 1/4 inch within any 32-inch measurement.

____ Foundation walls are not more than 1 inch out of level over the entire surface and do not vary more than 1/2 inch out of square when measured along the diagonal of a 6x8x10-foot triangle at any corner.

____ All line and pinholes are filled.

____ There are no honeycombs.

____ Ties are broken off and all tie holes filled.

____ Seams between the foundation and footing are sealed watertight.

____ Trench(es) is free and clear of all debris.

____ Anchor bolts are 6 feet on center and not more than 12 inches from corners.

____ All excess material is stacked and protected from weather.

____ Jobsite area is clean and free of debris, and excess concrete has been removed to driveway.

____ All formboards have been removed and stacked with excess material.

____ Field measurements match measurements on Trade Contractor's invoice.

Date: _____   Site Superintendent _____

Trade Contractor _____

<u>SCOPE OF WORK</u>
Standards and Description of Work Performance

ROOFING LABOR

The Company's **Terms and Conditions** are by reference a part of all **Scope of Work** requirements.

Construction Requirements:

Generally speaking the work of Trade Contractors and their employees is expected to be performed in a good and workmanlike manner. Workmanlike quality is defined as workmanship that meets or betters those criteria indicated in applicable building codes, using materials and installation methods identified in the construction plans and this Scope of Work.

Code Requirements:

All jobs shall conform to those standards stipulated in the building code, mechanical code, plumbing code, and electrical code applicable in the local jurisdiction. All construction on The Company's jobsites shall meet or exceed NAHB Performance and Building Standards.

General Comments:

The Company considers our Trade Contractors to be experts at producing a high-quality job. But everyone on our construction team—staff, Trade Contractors, and suppliers—must recognize the importance of providing quality in both the product and service areas while on our jobsites and in the homes of our purchasers.

Since we work as a team, poor quality or service, from any of us, reflects unfavorably on all of us. An exceptional level of product quality and highly effective service can help us all to increase our business and grow.

The Company's definition of quality construction also requires that every job be completed correctly the first time. When this does not occur it costs both of us additional money, imposes on the purchaser, and hurts our reputations as quality builders. That is why, in situations where construction was not completed in a quality manner, prompt corrective action is required to remedy specific deficiencies.

In the following information the term Site Superintendent shall refer to any The Company representative with authority to perform the specified task. The term Trade Contractor shall mean the Trade Contractor's organization or any representative that is assigned the authority to perform the specified task.

General: Roofing should be installed according to manufacturer's recommendations and should not leak under normal conditions.

Cut lines running up the roof on asphalt shingles should not vary more than 1/2 inch to either side of a line stretched from the eve to the ridge running parallel to the gable. Shingle edges running parallel to the

Company Rep's Initials _____
Trade Contractor's Initials _____

ridge should vary no more than 1/2 inch from a line parallel to the eve or ridge and stretched from one gable to the adjacent gable unless designed otherwise.

Shingle reveal at the ridge after capping shall be within 1 inch plus or minus of the intended shingle reveal stipulated by the manufacturer. Edge cuts shall be square and clean. The ridge cap shall begin at the roof end opposite the direction that prevailing winds travel. All nail holes shall be caulked with silicone caulk.

The Trade Contractors is required to use OSHA-approved safety measures for securing workers to the roof.

Installation Requirements: All work is to be done by trained, experienced individuals. Felt underlayment shall be installed over the entire roof as soon as the roof sheathing is complete. The underlayment shall lap at least 2 inches horizontally and 4 inches at any side lap, and should lap 6 inches at hips and ridges. Shingles should not be installed over wet underlayment or decking.

Roofing shall be applied only when the supporting roof decking is clean and dry. If any decking boards are warped the Site Superintendent should be notified immediately. Applying shingles over a bad decking job will result in a poor-quality job.

Roof sheathing shall not bow more than 1/2 inch in 2 feet.

Chimneys should be counterflashed to permit movement. A saddle should be installed where the chimney projects through the roof below the ridge.

Shingles should be secured using the correct number and size fasteners properly located on the shingle. Fasteners shall be compatible with the flashing used. All felt and shingles shall be installed per the manufacturer's instructions. The fasteners approved by the manufacturer must be used.

Safety: The Trade Contractor is the subcontractor most subject to serious injury on the jobsite. The Trade Contractor is required to have OSHA-approved ladders and use one of the approved safety devices that anchor roofers to the roof area. If pump-jack scaffolding is used by the Trade Contractor the scaffolding must have handrails installed. Toe boards are against code and shall not be used.

Turnkey Job: If the Trade Contractor has contracted for a turnkey job, the Trade Contractor shall furnish all materials of the type, brand, and color specified by The Company. All checks shall be made jointly to the roofing supplier and the Trade Contractor unless the Trade Contractor brings all materials onto the jobsite. If a supplier delivers roofing materials to the jobsite, then joint checks will be written and forwarded to the roofing supplier.

Warranty: The Trade Contractor is responsible for providing Warranty for any roof leaks for a period of one (1) year after the closing of the house. Other Warranty items and the time frame of coverage are found in The Company's printed Limited Warranty booklet.

The Trade Contractor is required to have an emergency number (beeper, etc.) that will allow him or her to be reached in case of an emergency on weekends, holidays, etc. This number shall be on a label and attached to the left door of the cabinet located to the left of the range.

_____ Company Rep's Initials
_____ Trade Contractor's Initials

The Trade Contractor shall have seven (7) days in which to correct any Warranty problem. If the problem is not corrected within seven (7) days then The Company shall correct the problem and will backcharge the Trade Contractor at the rate of $25.00 per hour, with a minimum charge of $100.00 plus the cost of any materials.

Inspection Reports: The Trade Contractor and the Site Superintendent shall walk the job together and complete each section of the inspection report(s). The Trade Contractor must correct any deficiency found during the inspection and the job must be 100-percent complete before payment will be made. The Trade Contractor and the Site Superintendent must sign-off on all sections of the inspection report(s) attesting that the job is 100-percent complete and is correct per the job requirements found in this Scope of Work.

Other: The Trade Contractor will be paid by the squares installed, not by the number of bundles delivered to the jobsite.

Detailed Job Requirements:

1. The Trade Contractor must walk the job and review the plans with the Site Superintendent before beginning work.

2. The Trade Contractor and the Site Superintendent must walk the job together and complete the pre-work section of the inspection report(s) before work may begin. The pre-work section of the inspection report(s) must be signed-off on by both parties.

3. Purchase orders should be picked up from the Site Superintendent.

4. One layer of nonperforated Type 15 felt underlayment is to be applied parallel to the eves lapping each course of the felt over the lower course at least 2 inches horizontally, 4 inches where ends join, and 6 inches at hips and ridges.

5. Felt is to be applied only to dry decking. It is not to be applied to damp, wet, or warped decking. Decking out of line more than 1/2 inch in 2 feet must be reported to the Site Superintendent for repair before felt is applied.

6. All asphalt shingles are to be applied per the installation instructions printed on each bundle of shingles. Shingles that slide out of position or fall off the roof are unacceptable.

7. All shingles shall be applied true and straight. A chalk line shall be used for coursing.

8. All shingles shall be blind fastened with all fasteners fully covered. Any nail holes shall be completely sealed.

9. If nails are used they must be corrosion-resistant nails, a minimum 12-gauge with a 3/8-inch head. If staples are used they must be corrosion-resistant staples, a minimum 16-gauge with a 15/16-inch crown width. All staples must be applied with pneumatic staplers.

10. Fasteners shall be long enough to penetrate into the decking a minimum of 1/4 inch, but shall not go all the way through the decking.

Company Rep's Initials _____
Trade Contractor's Initials _____

11. Valleys are to be constructed by centering 14-inch-wide aluminum valley metal flashing over the underlayment, securing with approved shingle fasteners, and covering with asphalt shingles.

12. Wall flashing shall be installed in the same manner as valley flashing.

13. All chimneys shall be flashed.

14. All roof vents (continuous ridge vents or turtle-back vents) shall be installed per plan.

15. A cover shingle is to be used to cover all horizontal and headwall flashing.

16. The Trade Contractor shall establish a drip edge to extend no less than 1/4 inch and not more than 3/4 inch over the facia.

17. All excess material shall be stacked in one location at the front of the lot. Excess material is to be counted, and amounts and types are to be listed on the purchase order.

18. All scrap material, shingle wrappers, and other trash shall be cleaned up and placed in the dumpster or the designated area for trash.

19. The Trade Contractor and Site Superintendent must walk the job together and perform a final inspection of the job. The final section of the inspection report(s) must be completed and signed off on. The inspection report(s) must be attached to the office's copy of the purchase order and the Trade Contractor's invoice or payment will not be issued.

20. Any items found during the final inspection that need correction shall be corrected before payment will be made.

I _____ agent for _____

_____ have read and fully understand the above **Scope of Work** and I hereby agree to perform all work in accordance with the above.

Date: _____   _____

Signed: Trade Contractor (or agent)

Date: _____   _____

For The Company

INSPECTION REPORT

ROOFING LABOR

Subdivision/Lot # _____

Pre- Work Inspection: (to be completed prior to beginning work)

____ Plans have been reviewed.

____ Purchase order has been picked up.

____ Work area is clean and free of debris.

____ All material is on site and ready to be used.

____ Decking is complete, dry, and there are no warped deck boards.

____ If windows are installed, all are intact and none are broken.

____ Trade Contractor has OSHA-approved equipment on site, with no toe boards.

Date: _____   Site Superintendent _____

Trade Contractor _____

Final Inspection: (to be completed before Trade Contractor leaves jobsite)

____ Roof is complete.

____ Shingles are straight and true.

____ Shingles are secure with no loose shingles.

____ Ridge vents (or turtle-back vents) are installed per plan.

____ Ridges are covered.

____ Drip edge is correct.

____ All valleys are completed and covered with shingles.

____ All nail holes are sealed.

____ All roof vents are installed per plan.

____ No shingle nails (or staples) protrude through the decking in the attic.

____ All excess material is stacked in one area, counted, and listed on the purchase order.

____ Jobsite is clean. All trash, debris, and shingle scraps have been removed.

____ Amount of squares installed is verified.

Date: _____   Site Superintendent _____

Trade Contractor _____

SCOPE OF WORK
Standards and Description of Work Performance

SHELVING

The Company's **Terms and Conditions** are by reference a part of all **Scope of Work** requirements.

Construction Requirements:

Generally speaking the work of Trade Contractors and their employees is expected to be performed in a good and workmanlike manner. Workmanlike quality is defined as workmanship that meets or betters those criteria indicated in applicable building codes, using materials and installation methods identified in the construction plans and this Scope of Work.

Code Requirements:

All jobs shall conform to those standards stipulated in the building code, mechanical code, plumbing code, and electrical code applicable in the local jurisdiction. All construction on The Company's jobsites shall meet or exceed NAHB Performance and Building Standards.

General Comments:

The Company considers our Trade Contractors to be experts at producing a high-quality job. But everyone on our construction team—staff, Trade Contractors, and suppliers—must recognize the importance of providing quality in both the product and service areas while on our jobsites and in the homes of our purchasers.

Since we work as a team, poor quality or service, from any of us, reflects unfavorably on all of us. An exceptional level of product quality and highly effective service can help us all to increase our business and grow.

The Company's definition of quality construction also requires that every job be completed correctly the first time. When this does not occur it costs both of us additional money, imposes on the purchaser, and hurts our reputations as quality builders. That is why, in situations where construction was not completed in a quality manner, prompt corrective action is required to remedy specific deficiencies.

In the following information the term Site Superintendent shall refer to any The Company representative with authority to perform the specified task. The term Trade Contractor shall mean the Trade Contractor's organization or any representative that is assigned the authority to perform the specified task.

General: Shelving will be of the type and size specified by The Company.

Installation: Installation shall be done by experienced, trained personnel. Shelving shall be installed to be plumb, level, and square, without bows, bends, or damage. All shelving shall be installed per the manufacturer's installation instructions and with the type of fasteners approved by the manufacturer.

_____ Company Rep's Initials
_____ Trade Contractor's Initials

Installation should ensure that the shelves do not damage drywall or other finished work.

Warranty: The Company believes that all work done and all materials installed in connection with one of our homes should be of high quality and that all Trade Contractors should stand behind the quality of their work and materials. Therefore we require all Trade Contractors to warrant the quality of their work for a period of one (1) year from date of closing of the house. Please refer to our printed Limited Warranty booklet for any specific items that are covered under Warranty as they apply to shelving.

The Trade Contractor shall have seven (7) days in which to correct any Warranty problem. If the problem is not corrected within seven (7) days then The Company shall correct the problem and will backcharge the Trade Contractor at the rate of $25.00 per hour, with a minimum charge of $100.00 plus the cost of any materials.

Inspection Reports: The Trade Contractor and the Site Superintendent shall walk the job together and complete each section of the inspection report(s). The Trade Contractor must correct any deficiency found during the inspection and the job must be 100-percent complete before payment will be made. The Trade Contractor and the Site Superintendent must sign-off on all sections of the inspection report(s) attesting that the job is 100-percent complete and is correct per the job requirements found in this Scope of Work.

Detailed Job Requirements:

1. Plans are subject to changes and modifications. It is the responsibility of the Trade Contractor to verify with the Site Superintendent that the plans have not changed before beginning work. The Trade Contractor at no cost to The Company will correct any errors that occur from using an incorrect set of plans.

2. Purchase orders should be picked up from the Site Superintendent.

3. The Trade Contractor and the Site Superintendent must walk the job together and complete the pre-work section of the inspection report(s) before work may begin. The pre-work section of the inspection report(s) must be signed-off on by both parties.

4. Shelving shall be a type specified by The Company. No substitutions are allowed.

5. All shelving shall be installed per the installation instructions of the manufacturer with the proper number of suggested fasteners.

6. All shelving shall be installed per plan.

7. All shelving shall be installed level, plumb, and square with no bows, bends, or damage to drywall or existing work.

8. House shall be left clean, broom-swept, and all debris shall be removed to the dumpster or to the designated trash site.

9. The Trade Contractor and Site Superintendent must walk the job together and perform a final inspection of the job. The final section of the inspection report(s) must be completed and

Company Rep's Initials _____
Trade Contractor's Initials _____

signed-off on by both parties. The inspection report must be attached to the office's copy of the purchase order and the Trade Contractor's invoice or payment will not be issued.

10. Any items found during the final inspection that need correction shall be corrected before payment will be made.

I _____ agent for _____

_____ have read and fully understand the above **Scope of Work** and I hereby agree to perform all work in accordance with the above.

Date: _____

_____
Signed: Trade Contractor (or agent)

Date: _____

_____
For The Company

INSPECTION REPORT

SHELVING

Subdivision/Lot # _____

Pre-Work Inspection: (to be completed prior to beginning work)

____   Purchase order has been received.

____   Plans have been verified.

____   Drywall and painting are complete and ready for shelving.

____   Drywall and paint are undamaged.

____   Windows are installed and undamaged.

____   House is clean, broom-swept, and free of debris.

Date: _____   Site Superintendent _____

                                          Trade Contractor _____

Final Inspection: (to be completed before Trade Contractor leaves jobsite)

____   All windows are intact and unbroken.

____   Drywall and paint is undamaged.

____   Shelving is installed level, plumb, square, and secure.

____   Shelving is installed per plan.

____   House is clean and broom-swept.

____   Debris has been removed to correct area.

Date: _____   Site Superintendent _____

                                          Trade Contractor _____

SCOPE OF WORK
Standards and Description of Work Performance

SHOWER DOORS

The Company's **Terms and Conditions** are by reference a part of all **Scope of Work** requirements.

Construction Requirements:

Generally speaking the work of Trade Contractors and their employees is expected to be performed in a good and workmanlike manner. Workmanlike quality is defined as workmanship that meets or betters those criteria indicated in applicable building codes, using materials and installation methods identified in the construction plans and this Scope of Work.

Code Requirements:

All jobs shall conform to those standards stipulated in the building code, mechanical code, plumbing code, and electrical code applicable in the local jurisdiction. All construction on The Company's jobsites shall meet or exceed NAHB Performance and Building Standards.

General Comments:

The Company considers our Trade Contractors to be experts at producing a high-quality job. But everyone on our construction team—staff, Trade Contractors, and suppliers—must recognize the importance of providing quality in both the product and service areas while on our jobsites and in the homes of our purchasers.

Since we work as a team, poor quality or service, from any of us, reflects unfavorably on all of us. An exceptional level of product quality and highly effective service can help us all to increase our business and grow.

The Company's definition of quality construction also requires that every job be completed correctly the first time. When this does not occur it costs both of us additional money, imposes on the purchaser, and hurts our reputations as quality builders. That is why, in situations where construction was not completed in a quality manner, prompt corrective action is required to remedy specific deficiencies.

In the following information the term Site Superintendent shall refer to any The Company representative with authority to perform the specified task. The term Trade Contractor shall mean the Trade Contractor's organization or any representative that is assigned the authority to perform the specified task.

General: The Company will determine the number, type, and size of all shower doors. No substitutions are allowed.

Installation: Installation shall be done by experienced, trained personnel. All shower doors shall be installed plumb, level, and square, without damage to the shower door, shower stall, or

_____ Company Rep's Initials
_____ Trade Contractor's Initials

surrounding areas. All shower doors shall be installed per the manufacturer's installation instructions and with the type of fasteners approved by the manufacturer.

Installation should ensure that there is no damage to drywall or other finished work. Shower doors should not leak under normal usage.

Warranty: The Company believes that all work done and all materials installed in connection with one of our homes should be of high quality and that all Trade Contractors should stand behind the quality of their work and materials. Therefore we require all Trade Contractors to warrant the quality of their work for a period of one (1) year from date of closing of the house. Please refer to our printed Limited Warranty booklet for any specific items that are covered under Warranty as they apply to shower doors.

The Trade Contractor shall have seven (7) days in which to correct any Warranty problem. If the problem is not corrected within seven (7) days then The Company shall correct the problem and will backcharge the Trade Contractor at the rate of $25.00 per hour, with a minimum charge of $100.00 plus the cost of any materials.

Inspection Reports: The Trade Contractor and the Site Superintendent shall walk the job together and complete each section of the inspection report(s). The Trade Contractor must correct any deficiency found during the inspection and the job must be 100-percent complete before payment will be made. The Trade Contractor and the Site Superintendent must sign-off on all sections of the inspection report(s) attesting that the job is 100-percent complete and is correct per the job requirements found in this Scope of Work.

Detailed Job Requirements:

1. Plans are subject to changes and modifications. It is the responsibility of the Trade Contractor to verify with the Site Superintendent that the plans have not changed before beginning work. Any errors that occur from using an incorrect set of plans will be corrected by the Trade Contractor at no cost to The Company.

2. The purchase order shall be mailed from the office. If a purchase order is not received prior to installation a copy of the purchase order should be picked up from the Site Superintendent.

3. The Trade Contractor and the Site Superintendent must walk the job together and complete the pre-work section of the inspection report(s) before work may begin. The pre-work section of the inspection report(s) must be signed-off on by both parties.

4. Shower doors shall be of the size, number, and type specified by The Company. No substitutions are allowed.

5. All shower doors shall be installed per the installation instructions of the manufacturer with the suggested fasteners.

6. All shower doors shall be installed per plan.

7. All shower doors shall be installed level, plumb, and square with no damage to doors, shower stall, or existing work.

Company Rep's Initials _____
Trade Contractor's Initials _____

8. Shower doors shall be installed so there will be no leakage during normal use.

9. House shall be left clean, broom-swept, and all debris shall be removed to the dumpster or to the designated trash site.

10. The Trade Contractor and Site Superintendent must walk the job together and perform a final inspection of the job. The final section of the inspection report(s) must be completed and signed-off on by both parties. The inspection report(s) must be attached to the office's copy of the purchase order and the Trade Contractor's invoice or payment will not be issued.

11. Any items found during the final inspection that need correction must be corrected before payment will be made.

I _____ agent for _____

_____ have read and fully understand the above **Scope of Work** and I hereby agree to perform all work in accordance with the above.

Date: _____

_____
Signed: Trade Contractor (or agent)

Date: _____

_____
For The Company

INSPECTION REPORT

SHOWER DOORS

Subdivision/Lot # _____

Pre-Work Inspection: (to be completed prior to beginning work)

____   Purchase order has been received.

____   Plans have been verified.

____   Shower is installed and ready for shower doors.

____   Adequate deadwood is installed.

____   Windows are installed and undamaged.

____   House is clean, broom-swept, and free of debris.

Date: _____   Site Superintendent _____

                                   Trade Contractor _____

Final Inspection: (to be completed before Trade Contractor leaves jobsite)

____   All windows are intact and unbroken.

____   Shower stall is undamaged.

____   Shower doors are installed level, plumb, and square.

____   Shower doors are installed per plan.

____   House is clean and broom-swept.

____   Debris has been removed to correct area.

Date: _____   Site Superintendent _____

                                   Trade Contractor _____

SCOPE OF WORK
Standards and Description of Work Performance

SHUTTERS

The Company's **Terms and Conditions** are by reference a part of all **Scope of Work** requirements.

Construction Requirements:

Generally speaking the work of Trade Contractors and their employees is expected to be performed in a good and workmanlike manner. Workmanlike quality is defined as workmanship that meets or betters those criteria indicated in applicable building codes, using materials and installation methods identified in the construction plans and this Scope of Work.

Code Requirements:

All jobs shall conform to those standards stipulated in the building code, mechanical code, plumbing code, and electrical code applicable in the local jurisdiction. All construction on The Company's jobsites shall meet or exceed NAHB Performance and Building Standards.

General Comments:

The Company considers our Trade Contractors to be experts at producing a high-quality job. But everyone on our construction team—staff, Trade Contractors, and suppliers—must recognize the importance of providing quality in both the product and service areas while on our jobsites and in the homes of our purchasers.

Since we work as a team, poor quality or service, from any of us, reflects unfavorably on all of us. An exceptional level of product quality and highly effective service can help us all to increase our business and grow.

The Company's definition of quality construction also requires that every job be completed correctly the first time. When this does not occur it costs both of us additional money, imposes on the purchaser, and hurts our reputations as quality builders. That is why, in situations where construction was not completed in a quality manner, prompt corrective action is required to remedy specific deficiencies.

In the following information the term Site Superintendent shall refer to any The Company representative with authority to perform the specified task. The term Trade Contractor shall mean the Trade Contractor's organization or any representative that is assigned the authority to perform the specified task.

General: The Company will specify the type of shutter to be used. No substitutions are allowed.

Installation: Installation shall be done by experienced, trained personnel. Shutters shall be installed plumb, level, and square, without bows, bends, or damage.

_____ Company Rep's Initials
_____ Trade Contractor's Initials

Warranty: The Company believes that all work done and all materials installed in connection with one of our homes should be of high quality and that all Trade Contractors should stand behind the quality of their work and materials. Therefore we require all Trade Contractors to warrant the quality of their work for a period of one (1) year from date of closing of the house. Please refer to our printed Limited Warranty booklet for any specific items that are covered under Warranty as they apply to shutters.

The Trade Contractor shall have seven (7) days in which to correct any Warranty problem. If the problem is not corrected within seven (7) days then The Company shall correct the problem and will backcharge the Trade Contractor at the rate of $25.00 per hour, with a minimum charge of $100.00 plus the cost of any materials.

Inspection Reports: The Trade Contractor and the Site Superintendent shall walk the job together and complete each section of the inspection report(s). The Trade Contractor must correct any deficiency found during the inspection and the job must be 100-percent complete before payment will be made. The Trade Contractor and the Site Superintendent must sign-off on all sections of the inspection report(s) attesting that the job is 100-percent complete and is correct per the job requirements found in this Scope of Work.

Detailed Job Requirements:

1. The Trade Contractor must check in with the Site Superintendent prior to beginning work.

2. The purchase order shall be mailed from the office. If a purchase order has not been received prior to installation a copy of the purchase order should be picked up from the Site Superintendent.

3. The Trade Contractor and the Site Superintendent must walk the job together and complete the pre-work section of the inspection report(s) before work may begin. The pre-work section of the inspection report(s) must be signed-off on by both parties.

4. Shutters shall be the type specified by The Company. No substitutions are allowed.

5. The Site Superintendent will notify the Trade Contractor of the correct type, color, and number of shutters to be installed in adequate time for the Trade Contractor to secure the correct shutters.

6. All shutters shall be installed per the installation instructions and with the manufacturer's suggested fasteners.

7. All shutters shall be installed per plan.

8. No shutter shall be installed that has visible defects such as splits, unpainted areas, or out-of-plumb shape, etc. Any shutter with either a horizontal or vertical deflection of more than 1/2 inch when measured from top to bottom or side to side is unacceptable.

9. All shutters shall be installed level, plumb, and square with no bows, bends, or damage. Shutters installed on uneven surfaces, such as stone, shall be secured to the structure so that the shutters do not "wave" over the surface.

Company Rep's Initials _____
Trade Contractor's Initials _____

10. The Trade Contractor and Site Superintendent must walk the job together and perform a final inspection of the job. The final section of the inspection report(s) must be completed and signed-off on by both parties. The inspection report(s) must be attached to the office's copy of the purchase order and the Trade Contractor's invoice or payment will not be issued.

11. Any items found during the final inspection that need correction shall be corrected before payment will be made.

I _____ agent for _____

_____ have read and fully understand the above **Scope of Work** and I hereby agree to perform all work in accordance with the above.

Date: _____

_____
Signed: Trade Contractor (or agent)

Date: _____

_____
For The Company

INSPECTION REPORT

SHUTTERS

Subdivision/Lot # _____

Pre-Work Inspection: (to be completed prior to beginning work)

____   Purchase order has been received.

____   Shutters are correct type, size, and color.

____   Siding, brick, stucco, etc., are complete.

____   Windows are installed and undamaged.

Date: _____   Site Superintendent _____

Trade Contractor _____

Final Inspection: (to be completed before Trade Contractor leaves jobsite)

____   All windows are intact and unbroken.

____   Siding, brick, and stucco are undamaged.

____   Shutters are installed level, plumb, and square with no bows or bends.

____   Shutters are undamaged.

____   If landscaping is installed, there has been no damage to landscaping.

Date: _____   Site Superintendent _____

Trade Contractor _____

SCOPE OF WORK
Standards and Description of Work Performance

SIDING and CORNICE LABOR

The Company's **Terms and Conditions** are by reference a part of all **Scope of Work** requirements.

Construction Requirements:

Generally speaking the work of Trade Contractors and their employees is expected to be performed in a good and workmanlike manner. Workmanlike quality is defined as workmanship that meets or betters those criteria indicated in applicable building codes, using materials and installation methods identified in the construction plans and this Scope of Work.

Code Requirements:

All jobs shall conform to those standards stipulated in the building code, mechanical code, plumbing code, and electrical code applicable in the local jurisdiction. All construction on The Company's jobsites shall meet or exceed NAHB Performance and Building Standards.

General Comments:

The Company considers our Trade Contractors to be experts at producing a high-quality job. But everyone on our construction team—staff, Trade Contractors, and suppliers—must recognize the importance of providing quality in both the product and service areas while on our jobsites and in the homes of our purchasers.

Since we work as a team, poor quality or service, from any of us, reflects unfavorably on all of us. An exceptional level of product quality and highly effective service can help us all to increase our business and grow.

The Company's definition of quality construction also requires that every job be completed correctly the first time. When this does not occur it costs both of us additional money, imposes on the purchaser, and hurts our reputations as quality builders. That is why, in situations where construction was not completed in a quality manner, prompt corrective action is required to remedy specific deficiencies.

In the following information the term Site Superintendent shall refer to any The Company representative with authority to perform the specified task. The term Trade Contractor shall mean the Trade Contractor's organization or any representative that is assigned the authority to perform the specified task.

General: The type of siding used on a house is determined by the price range, subdivision, and options selected by the purchaser.

Material: Concrete siding and vinyl siding are the normal siding used. The purchase order will specify the type of siding to be used. The Company will furnish the siding material. The Trade Contractor will furnish all fasteners.

_____ Company Rep's Initials
_____ Trade Contractor's Initials

Installation: All siding will be installed per the manufacturer's specifications. All vinyl siding will be installed by a Trade Contractor trained and certified in the installation of vinyl siding.

All siding shall be installed by trained, experienced individuals. All siding shall be installed per plan using the fasteners required by the manufacturer. Siding is to be nailed, blind or face, per the manufacturer's installation instructions.

The Trade Contractor is expected to eliminate excess waste of materials.

No siding shall be installed over any area not covered with insulation board or over broken or badly damaged insulation board.

Safety: The Trade Contractor shall follow all safety rules in regards to ladders, scaffolding, safety rails, etc. If pump-jack-type scaffolding is used it must include handrails.

Warranty: The Company believes that all work done in connection with one of our homes should be of high quality and that all Trade Contractors should stand behind the quality of their work. Therefore we require all Trade Contractors to warrant the quality of their work for a period of one (1) year from date of closing of the house. Please refer to our printed Limited Warranty booklet for any specific items that are covered under Warranty as they apply to siding and cornices.

The Trade Contractor shall have seven (7) days in which to correct any Warranty problem. If the problem is not corrected within seven (7) days then The Company shall correct the problem and will backcharge the Trade Contractor at the rate of $25.00 per hour, with a minimum charge of $100.00 plus the cost of any materials.

Inspection Reports: The Trade Contractor and the Site Superintendent shall walk the job together and complete each section of the inspection report(s). The Trade Contractor must correct any deficiency found during the inspection and the job must be 100-percent complete before payment will be made. The Trade Contractor and the Site Superintendent must sign-off on all sections of the inspection report(s) attesting that the job is 100-percent complete and is correct per the job requirements found in this Scope of Work.

Detailed Job Requirements:

1. The Trade Contractor shall review the plans with the Site Superintendent prior to beginning work to ensure that the Trade Contractor is familiar with the work to be done and the materials to be used. Any corrections that have to be made because the Trade Contractor did not review the plans with the Site Superintendent shall be corrected at the expense of the Trade Contractor.

2. Purchase orders should be picked up from the Site Superintendent.

3. The Trade Contractor and the Site Superintendent must walk the job together and complete the pre-work section of the inspection report(s) before work may begin. The pre-work section of the inspection report(s) must be signed-off on by both parties.

4. All siding shall be installed per the manufacturer's installation instructions using the fasteners and fastening procedures required by the manufacturer.

Company Rep's Initials _____
Trade Contractor's Initials _____

5. All siding shall be level and straight with a maximum of 1/2 inch off parallel in 20 feet with contiguous courses.

6. End gaps wider than 3/16 inch are unacceptable in hardboard or concrete siding.

7. All siding shall be secured to solid framing.

8. All siding shall be installed over insulation boards. Siding may not be installed in any area missing insulation boards or over any broken or badly damaged insulation boards. Insulation boards that are broken must be repaired or replaced.

9. Any siding board that is cracked, split, or broken during installation must be removed and replaced. No broken, split, or cracked siding boards shall be on the house at the completion of the job.

10. Siding waves in vinyl siding of more than 1/4 inch in 16 inches is unacceptable.

11. Bows in hardboard or concrete siding of more than 1/2 inch in 32 inches are unacceptable.

12. "Z" mold flashing shall be installed at the top of all openings. Additional flashing shall be installed as required.

13. Facia, rake boards, associated soffits, and soffit areas are the responsibility of the roofing Trade Contractor.

14. Facia, rake boards, and soffit material should not be installed if split, broken, gouged, or nicked, or if it contains large knots, holes or is badly machined.

15. All exterior trim shall be installed per plan, including any porch columns and railings. Exterior trim should not be installed if split, broken, or gouged, or if it contains large knots or holes. No exterior trim of poor appearance shall be used. Splits wider than 1/8 inch are unacceptable. Bows and twists exceeding 3/8 of an inch in 8 feet are unacceptable.

16. Joints between exterior trim elements shall not result in joints opened wider than 1/4 inch. In all cases, the exterior trim shall perform its function of excluding the elements.

17. Exterior trim shall be tightly secured to framing members and shall be straight, level, and plumb.

18. All inside and outside corners and window and door openings shall be trimmed out per plan.

19. Garage doorjambs shall be completed to the size of the garage door as shown on the plans. Corners shall be completed per plans.

20. Garage doorjambs shall be square, level, plumb, and straight.

21. Any holes, nicks, gouges, etc., in siding shall be repaired before job is considered complete.

22. Unacceptable pieces of siding shall be replaced before the job is considered complete.

_____ Company Rep's Initials
_____ Trade Contractor's Initials

23. All construction debris is to be removed to the dumpster or designated trash collection area and the jobsite is to be left clean.

24. The Trade Contractor and Site Superintendent must walk the job together and perform a final inspection of the job. The final section of the inspection report(s) must be completed and signed-off on by both parties. The inspection report(s) must be attached to the office's copy of the purchase order and the Trade Contractor's invoice or payment will not be issued.

25. Any items found during the final inspection that need correction shall be corrected before payment will be made.

I _____ agent for _____

_____ have read and fully understand the above **Scope of Work** and I hereby agree to perform all work in accordance with the above.

Date: _____

_____
Signed: Trade Contractor (or agent)

Date: _____

_____
For The Company

INSPECTION REPORT

SIDING AND CORNICE LABOR

Subdivision/Lot # _____

Pre-Work Inspection: (to be completed prior to beginning work)

____   Plans have been reviewed with the Site Superintendent.

____   Purchase order has been received.

____   Materials are on site and ready to be used.

____   Area around house is free and clear of debris.

____   All insulated boards are installed with no open areas or gaps and no broken boards.

____   Insulation boards are dry.

____   Roof is completed.

____   Windows and doors are set.

____   Windows are intact and unbroken.

Date: _____   Site Superintendent _____

                             Trade Contractor _____

Final Inspection: (to be completed before Trade Contractor leaves jobsite)

____   From interior of house all fasteners are in framing members. Any area where fasteners have missed framing members and are protruding through insulation boards must be refastened.

____   All broken insulation boards have been repaired and sealed.

____   Siding is installed per manufacturer's installation instructions.

____   Siding is level and straight with no more than a 1/4-inch deviation in 10 feet.

____   Exterior trim is level, plumb, and undamaged.

____   Exterior trim is not split or broken, and does not have large knots or holes.

____   Exterior trim is secured tightly to framing members.

____   Gaps in exterior trim do not exceed 1/4 inch and have been caulked.

____   Facia, rake boards, soffits, and soffit areas are all complete. Material is undamaged with no holes, splits, broken boards, large knots, etc.

____   No siding boards are broken.

____   All holes, nicks, gouges, etc., are caulked and/or repaired.

____   Site is clean and free of debris.

____   Excess material is stacked at front of lot.

Date: _____   Site Superintendent _____

                             Trade Contractor _____

<u>SCOPE OF WORK</u>
Standards and Description of Work Performance

SLABS

The Company's **<u>Terms and Conditions</u>** are by reference a part of all **<u>Scope of Work</u>** requirements.

<u>Construction Requirements</u>:

Generally speaking the work of Trade Contractors and their employees is expected to be performed in a good and workmanlike manner. Workmanlike quality is defined as workmanship that meets or betters those criteria indicated in applicable building codes, using materials and installation methods identified in the construction plans and this Scope of Work.

<u>Code Requirements</u>:

All jobs shall conform to those standards stipulated in the building code, mechanical code, plumbing code, and electrical code applicable in the local jurisdiction. All construction on The Company's jobsites shall meet or exceed NAHB Performance and Building Standards.

<u>General Comments</u>:

The Company considers our Trade Contractors to be experts at producing a high-quality job. But everyone on our construction team—staff, Trade Contractors, and suppliers—must recognize the importance of providing quality in both the product and service areas while on our jobsites and in the homes of our purchasers.

Since we work as a team, poor quality or service, from any of us, reflects unfavorably on all of us. An exceptional level of product quality and highly effective service can help us all to increase our business and grow.

The Company's definition of quality construction also requires that every job be completed correctly the first time. When this does not occur it costs both of us additional money, imposes on the purchaser, and hurts our reputations as quality builders. That is why, in situations where construction was not completed in a quality manner, prompt corrective action is required to remedy specific deficiencies.

In the following information the term Site Superintendent shall refer to any The Company representative with authority to perform the specified task. The term Trade Contractor shall mean the Trade Contractor's organization or any representative that is assigned the authority to perform the specified task.

<u>General</u>: Concrete is subject to various phenomena including applied chemicals and natural elements that can deteriorate the surface. In some cases however, surface defects such as flaking, scaling, or spallling may be caused by improper finishing. Excessive powdering or chalking may occur due to improper troweling, excess water, or when uncured concrete freezes. It is important that proper precautions and correct techniques be utilized in the handling and finishing of concrete in the field to ensure a quality job.

Company Rep's Initials _____
Trade Contractor's Initials _____

Poured concrete should not crack in excess of 3/16 inch in width or vertical displacement. Concrete shall not pit, scale, or spall to the extent that the aggregate is exposed under normal weathering and use.

Concrete floors in rooms designed as living space shall have no displacement, no visible cracks, nor any cracks that will damage finish flooring, nor any crack in excess of 3/16 inch when fully cured. These areas shall have no pits, depressions, or areas of unevenness exceeding 1/4 inch in 32 inches.

Walks, stoops, steps, patios, and garage floors shall not heave, settle, or separate from the main house in excess of 1/2 inch. Outdoor walks, stoops, steps, and patios should be finished so that water drains off.

All patios, stoops, and garages shall be poured at the same time the slab is poured. Drives and walks are poured at a later date. All slabs shall be 3,000 psi. Garages, walks, patios, and stoops shall be 2,500 psi.

It is the Trade Contractor's responsibility to ensure that no damage is done to any other Trade Contractor's pipe or material placed in the slab prior to and during the placement of the concrete.

Handling Requirements: All work is to be done by trained, experienced individuals. Forms should be tight and well braced. They should be moistened in hot weather to prevent water extraction. Snow and ice must be removed from forms prior to pouring.

Concrete should be placed at an appropriate rate so that it can be spread, straightened, and darbied properly. Techniques for handling and placing concrete should ensure that it remains uniform within each batch and from batch to batch. Concrete should not be allowed to run or be worked over long distances, and should not be allowed to drop more than 4 feet.

To avoid water collection, placement should start at ends working toward center on walls, and should begin around perimeter on slabs.

Finishing Requirements: Concrete should be placed as soon and as continuously as possible. Then excess concrete should be screeded off. To avoid walking into screeded area, screed stakes should be removed as the work progresses.

Screeding should be followed immediately by bringing mortar to the true surface grade once by darbying. Darbying should embed coarse aggregate and eliminate voids and ridges left by screeding.

When the concrete has lost its sheen and begins to stiffen, control joints may be cut (if required) and the slab may be edged. The concrete surface should then be floated to remove high and low spots and surface imperfections left by edging or jointing, and to consolidate the mortar at the surface.

The surface then shall be immediately steel troweled to a smooth hard surface. Where a smooth surface is not desirable (walks, stoops, drives, and patios) the surface should be brushed lightly to a nonslip surface (broom-finished).

Curing Requirements: To ensure maximum strength, concrete should be protected from rapid drying by covering with polyethylene. Forms should be left in place as long as practical. The surface should be kept uniformly wet or moist through the curing period. In cold weather with air

_____ Company Rep's Initials
_____ Trade Contractor's Initials

temperatures below 40 degrees Fahrenheit (F), the water and aggregate should be heated so that the mixed concrete is placed at a temperature between 50 and 70 degrees F.

For monolithic slabs, forms should remain intact until concrete has cured sufficiently to ensure structural stability. The concrete should be hard enough that form removal with reasonable care causes no damage to the surface finish.

Warranty: The Company believes that all work done in connection with one of our homes should be of high quality and that all Trade Contractors should stand behind the quality of their work. Therefore we require all Trade Contractors to warrant the quality of their work for a period of one (1) year from date of closing of the house. Please refer to our printed Limited Warranty booklet for any specific items that are covered under Warranty as they apply to the concrete slab.

The Trade Contractor shall have seven (7) days in which to correct any Warranty problem. If the problem is not corrected within seven (7) days then The Company shall correct the problem and will backcharge the Trade Contractor at the rate of $25.00 per hour, with a minimum charge of $100.00 plus the cost of any materials.

Inspection Reports: The Trade Contractor and the Site Superintendent shall walk the job together and complete each section of the inspection report(s). The Trade Contractor must correct any deficiency found during the inspection and the job must be 100-percent complete before payment will be made. The Trade Contractor and the Site Superintendent must sign-off on all sections of the inspection report(s) attesting that the job is 100-percent complete and is correct per the job requirements found in this Scope of Work.

Detailed Job Requirements:

1. A new set of plans is required for each house. Plans are subject to changes and modifications. It is the responsibility of the Trade Contractor to have new plans before beginning work. Plans should be picked up at the job trailer from the Site Superintendent. The Trade Contractor at no cost to The Company will correct any errors that occur from using an incorrect set of plans.

2. Purchase orders should be picked up from the Site Superintendent.

3. The Trade Contractor and the Site Superintendent must walk the job together and complete the pre-work section of the inspection report(s) before work may begin. The pre-work section of the inspection report(s) must be signed-off on by both parties.

4. The Trade Contractor shall set forms.

5. The Trade Contractor shall fine-grade for the slab. The area within the foundation walls shall be free of vegetation and foreign material.

6. The Trade Contractor shall fill the floor area. The fill shall be compacted to assure uniform support of the slab. Fill depths shall not exceed 24 inches. Any fill over 24 inches requires an engineer's compression report.

Company Rep's Initials _____
Trade Contractor's Initials _____

7. The Trade Contractor shall hand-dig grade beams (if required), fill forms, and set two (2) pieces of #4 reinforcing steel in grade beams.

8. The Trade Contractor shall spread a minimum of 4 inches of 57-stone in slab area.

9. The Trade Contractor shall place polyethylene in the slab. Polyethylene shall be lapped no less than 6 inches and shall be placed between the concrete floor slab and the base course.

10. The Trade Contractor shall place wire mesh reinforcement of 6x6x10 WWL that extends down into the footings. The mesh reinforcement shall be pulled up to the mid-point of the slab when the concrete is poured to ensure its maximum efficiency within the footing.

11. The Trade Contractor shall pour and finish concrete slabs. Concrete in all slabs shall be 3,000 psi.

12. The Trade Contractor shall steel-trowel finish all slabs.

13. The Trade Contractor shall broom-finish all garages, walks, drives, patios, and stoops, all of which shall be 2,500 psi.

14. Slab areas of unevenness, pits, or depressions exceeding 1/4 inch in 32 inches are unacceptable.

15. All edges above grade shall be smooth with no aggregate exposed, and shall be rubbed accordingly.

16. The Trade Contractor shall furnish all equipment and labor to ensure all fill areas at masonry and/or concrete wall have a minimum of 95-percent compaction.

17. The Trade Contractor shall supply rough plumbing backfill and adequate subslab compaction.

18. Upon notification by the Site Superintendent, all piers, column bases, and stem walls used to accommodate nontypical site conditions shall be dug, formed, and poured.

19. Patios shall be poured either monolithically with the building slab or separately using proper expansion joints at the connection areas. In either case, patio slabs shall be 1-1/2 inches lower than the finished floor of the base building and 1/4-inch slope per 1 foot to the outside. Patios shall be poured 4 inches in depth.

20. All garage floors, patios, and front stoops or porches shall be poured simultaneously with the slab. All garage floors must be sloped per code toward the garage doors.

21. The trench at that footing/foundation area shall be free of material and debris.

22. All trash and debris shall be removed from lot to dumpster or the designated area for trash.

23. All runoff concrete shall be removed to the driveway cut.

24. Any concrete spilled, splashed, etc., on foundation shall be cleaned while still wet with no damage to foundation.

_____ Company Rep's Initials
_____ Trade Contractor's Initials

25. Slab, garage floors, patios, and front stoops or porches shall be field-measured and the actual square footage shall be listed on the Trade Contractor's copy of the purchase order and on the Trade Contractor's invoice.

26. The Trade Contractor is responsible for cleaning up all residual materials before the job shall be accepted as complete.

27. Any additional work required because of a nontypical lot (piers, column bases, etc.) will be considered as a "change order" and will be paid per the price specified on the change order. The change order must be approved by The Company's purchasing agent. Any additional work must be submitted on a separate invoice and must include the change order/purchase order number(s).

28. The Trade Contractor and Site Superintendent must walk the job together and perform a final inspection of the job. The final section of the inspection report(s) must be completed and signed-off on by both parties. The inspection report(s) must be attached to the office's copy of the purchase order and the Trade Contractor's invoice or payment will not be issued.

29. Any items found during the final inspection that need correction shall be corrected before payment will be made.

I _____ agent for _____

_____ have read and fully understand the above **Scope of Work** and I hereby agree to perform all work in accordance with the above.

Date: _____   _____
                            Signed: Trade Contractor (or agent)

Date: _____   _____
                            For The Company

INSPECTION REPORT

SLABS

Subdivision/Lot # _____

Pre-Work Inspection: (to be completed prior to beginning work)

____ Plans have been picked up and signed for.

____ Purchase order has been picked up.

____ Work area is clean and free of debris.

____ Foundation is complete, aligned, and corners are square.

____ Seam between foundation and footing is sealed watertight.

____ Dimensions are correct.

____ Batterboards have been removed.

____ Straightness (alignment) of lines have been verified.

____ All materials are on site and ready to be used.

____ Clear access and a stable base is available for concrete trucks.

Date: _____   Site Superintendent _____

Trade Contractor _____

Final Inspection: (to be completed before Trade Contractor leaves jobsite)

____ Slab is level and plumb with no pits, depressions, or areas of unevenness exceeding 1/4 inch in 32 inches.

____ Slab surface is trowel-finished and smooth.

____ Garage floor is poured and broom-finished.

____ Garage floor slopes toward door per code with no pits, depressions, or areas of unevenness exceeding 1/4 inch in 32 inches.

____ Patio, stoop, and porch are poured and broom-finished with correct slope for drainage.

____ Patio, stoop, and porch are no more than 1-1/2 inches from the house slab.

____ No exposed aggregate is in any area.

____ No cracks or displacement are in any area.

____ Excess 57-stone is spread in driveway cut.

____ Excess concrete runoff has been moved to driveway cut.

____ Jobsite is clean and free of debris.

____ Foundation trench is free and clear of debris.

____ Foundation is clean of any concrete spills, overruns, etc.

Date: _____   Site Superintendent _____

Trade Contractor _____

SCOPE OF WORK
Standards and Description of Work Performance

STAIRS (PREBUILT)

The Company's **Terms and Conditions** are by reference a part of all **Scope of Work** requirements.

Construction Requirements:

Generally speaking the work of Trade Contractors and their employees is expected to be performed in a good and workmanlike manner. Workmanlike quality is defined as workmanship that meets or betters those criteria indicated in applicable building codes, using materials and installation methods identified in the construction plans and this Scope of Work.

Code Requirements:

All jobs shall conform to those standards stipulated in the building code, mechanical code, plumbing code, and electrical code applicable in the local jurisdiction. All construction on The Company's jobsites shall meet or exceed NAHB Performance and Building Standards.

General Comments:

The Company considers our Trade Contractors to be experts at producing a high-quality job. But everyone on our construction team—staff, Trade Contractors, and suppliers—must recognize the importance of providing quality in both the product and service areas while on our jobsites and in the homes of our purchasers.

Since we work as a team, poor quality or service, from any of us, reflects unfavorably on all of us. An exceptional level of product quality and highly effective service can help us all to increase our business and grow.

The Company's definition of quality construction also requires that every job be completed correctly the first time. When this does not occur it costs both of us additional money, imposes on the purchaser, and hurts our reputations as quality builders. That is why, in situations where construction was not completed in a quality manner, prompt corrective action is required to remedy specific deficiencies.

In the following information the term Site Superintendent shall refer to any The Company representative with authority to perform the specified task. The term Trade Contractor shall mean the Trade Contractor's organization or any representative that is assigned the authority to perform the specified task.

General: When risers are closed, all treads may have a uniform projection not to exceed 1-1/2 inches. The greatest height within any flight of stairs shall not exceed the smallest by more than 3/8 inch. The greatest tread run within any flight of stairs shall not exceed the smallest by more than 3/8 inch. Stairways shall not be less than 3 feet in clear width.

Company Rep's Initials _____
Trade Contractor's Initials _____

The Trade Contractor is responsible for furnishing temporary stairs. Temporary stairs are to be delivered to the jobsite on an "as-needed" basis. The Trade Contractor will be notified by the Site Superintendent when temporary stairs are needed. The framing Trade Contractor will be responsible for installing the temporary stairs. When the permanent stairs are delivered by the Trade Contractor, the Trade Contractor shall remove the temporary stairs and place them in the garage to be used on the next house.

Installation: Due care must be taken when removing temporary stairs and installing the permanent stairs. The installers must not damage the existing framing or other Trade Contractors' work. All stairs are to be properly braced on the underside of the stairs to prevent any movement, including bouncing, creaks, squeaks, and movement. All "shims and bracing pieces" on the underside of treads must be secured with both additional shims and nails.

All stairs shall be covered with rosin paper to protect the stair treads during the remaining time of construction. All areas of oak on the stair treads are to be doubly protected with a material that cannot be readily torn, knocked off, or otherwise removed to leave the oak exposed to damage.

Warranty: The Company believes that all work done and all materials installed in connection with one of our homes should be of high quality and that all Trade Contractors should stand behind the quality of their work and materials. Therefore we require all Trade Contractors to warrant the quality of their work for a period of one (1) year from date of closing of the house. Please refer to our printed Limited Warranty booklet for any specific items that are covered under Warranty as they apply to stairs.

The Trade Contractor shall have seven (7) days in which to correct any Warranty problem. If the problem is not corrected within seven (7) days then The Company shall correct the problem and will backcharge the Trade Contractor at the rate of $25.00 per hour, with a minimum charge of $100.00 plus the cost of any materials.

Inspection Reports: The Trade Contractor and the Site Superintendent shall walk the job together and complete each section of the inspection report(s). The Trade Contractor must correct any deficiency found during the inspection and the job must be 100-percent complete before payment will be made. The Trade Contractor and the Site Superintendent must sign-off on all sections of the inspection report(s) attesting that the job is 100-percent complete and is correct per the job requirements found in this Scope of Work.

Detailed Job Requirements:

1. Prior to beginning work the Site Superintendent shall review the plans with the Trade Contractor to ensure the accuracy of the job.

2. The purchase order will be mailed to the Trade Contractor.

3. The Trade Contractor and the Site Superintendent must walk the job together and complete the pre-work section of the inspection report(s) before work may begin. The pre-work section of the inspection report(s) must be signed-off on by both parties.

4. The Trade Contractor shall furnish temporary stairs as required.

_____ Company Rep's Initials
_____ Trade Contractor's Initials

5. The Trade Contractor shall remove temporary stairs and place them in the garage when permanent stairs are installed.

6. The Trade Contractor shall construct stairs in such a manner that stairs do not squeak, creak, bounce, or move.

7. The Trade Contractor shall field-measure the stair area to ensure a correct fit and riser spans.

8. Stairs are to be fitted to the floor correctly for the floor covering used (wood, vinyl, or carpet).

9. All stair treads, especially all oak areas of the tread, shall be covered and protected.

10. The Trade Contractor shall add additional bracing as required to ensure the stability of the stairs.

11. The Trade Contractor shall prevent damage to existing framing or other completed construction.

12. Any additional work required because of additional stairs will be considered a change order and will be paid per the price specified on the change order/purchase order. Any additional work must be submitted on a separate invoice and must include the change order/purchase order number.

13. The Trade Contractor and Site Superintendent must walk the job together and perform a final inspection of the job. The final section of the inspection report(s) must be completed and signed-off on by both parties. The inspection report(s) must be attached to the office's copy of the purchase order and the Trade Contractor's invoice or payment will not be issued.

14. The Trade Contractor is responsible for cleaning up all residual materials before the job shall be accepted as complete.

15. Any items found during the final inspection that need correction shall be corrected before payment will be made.

I _____ agent for _____

_____ have read and fully understand the above **Scope of Work** and I hereby agree to perform all work in accordance with the above.

Date: _____   _____
                            Signed: Trade Contractor (or agent)

Date: _____   _____
                            For The Company

INSPECTION REPORT

STAIRS (PREBUILT)

Subdivision/Lot # _____

Pre-Work Inspection:

____ Plans have been reviewed with Site Superintendent.

____ Purchase order has been received.

____ House has been field-measured.

____ Work area is clear of debris.

Date: _____ Site Superintendent _____

Trade Contractor _____

Final Inspection: (to be completed before Subcontractor leaves jobsite)

____ Stairs are installed in levels that require stairs.

____ All sets of stairs fit correctly.

____ All sets of stairs are the correct type.

____ Heights and runs have been measured and are correct for all sets of stairs.

____ All sets of stairs do not creak, squeak, or move.

____ Bracing has been checked (nailed and glued) and verified for all sets of stairs.

____ No patching has been made to any stairs due to incorrect measurements.

____ Temporary stairs have been moved to garage.

____ All risers are protected on all sets of stairs.

____ All oak has double protection. No oak is visible.

____ If extra stairs are required for the garage they are the correct height and are correctly installed.

Date: _____ Site Superintendent _____

Trade Contractor _____

SCOPE OF WORK
Standards and Description of Work Performance

TRIM LABOR

The Company's **Terms and Conditions** are by reference a part of all **Scope of Work** requirements.

Construction Requirements:

Generally speaking, the work of Trade Contractors and their employees is expected to be performed in a good and workmanlike manner. Workmanlike quality is defined as workmanship which meets or betters those criteria indicated in applicable building codes, using materials and installation methods identified in the construction plans and this Scope of Work.

Code Requirements:

All jobs shall conform to those standards stipulated in the Building code, Mechanical Code, Plumbing Code and Electrical Code applicable in the local jurisdiction. All construction on The Company's job sites shall meet or exceed NAHB Performance and Building Standards.

General Comments:

The Company considers our Trade Contractors to be experts at producing a high-quality job. But everyone on our construction team—staff, Trade Contractors, and suppliers—must recognize the importance of providing quality in both the product and service areas while on our jobsites and in the homes of our purchasers.

Since we work as a team, poor quality or service from any of us reflects unfavorably upon all of us. An exceptional level of product quality and highly effective service can help us all to increase our business and grow.

The Company's definition of quality construction also requires that every job be completed correctly the first time. When this does not occur it costs both of us additional money, imposes on the purchaser and hurts our reputations as quality builders. That is why, in situations where construction was not completed in a quality manner, prompt corrective action is required to remedy specific deficiencies.

In the following information the term Site Superintendent shall refer to any The Company representative with authority to perform the specified task. The term Trade Contractor shall mean the Trade Contractor's organization or any representative that is assigned the authority to perform the specified task.

General: For most buyers, the look of finished carpentry, millwork, and cabinetry is most critical in their judgement of quality construction. Care should be taken in the storage and handling of finish materials to avoid damage and soiling. Installed materials should be protected when necessary.

Company Rep's Initials _____
Trade Contractor's Initials _____

Materials: Extra care should be taken to fully inspect trim materials for damage and unacceptable trim. Interior doors are delivered prehung and should be stored flat on a level surface in a clean, dry, well-ventilated location. Warped doors should not be installed.

Installation: All work is to be done by trained, experienced individuals. Trim and molding should be installed tightly to the wall surface and fastened securely. Joints in moldings and adjacent surfaces should not exceed 1/8 inch in width.

Doors should be installed so that they operate properly as intended and with reasonable ease. Interior doors should be installed level, plumb in both directions, and squarely in the opening with no more than 1/4 inch in 4 feet deviation in any direction. Prehung doors should be checked for jamb squareness and straightness. Reinstall manufacturer's braces to square units where necessary. After leveling and plumbing the unit, shim and nail securely. All bifold doors shall hang squarely and open smoothly.

Warranty: The Company believes that all work done and all materials installed in connection with one of our homes should be of high quality and that all Trade Contractors should stand behind the quality of their work. Therefore we require all Trade Contractors to warrant the quality of their work for a period of one (1) year from date of closing of the house. Please refer to our printed Limited Warranty booklet for specific items that are covered under Warranty as they apply to trim.

The Trade Contractor shall have seven (7) days in which to correct any warranty problem. If the problem is not corrected within seven (7) days then The Company shall correct the problem and will backcharge the Trade Contractor at the rate of $25.00 per hour, with a minimum charge of $100.00 plus the cost of any materials.

Inspection Reports: The Trade Contractor and the Site Superintendent shall walk the job together and complete each section of the inspection report(s). The Trade Contractor must correct any deficiency found during the inspection and the job must be 100-percent complete before payment will be made. The Trade Contractor and the Site Superintendent must sign-off on all sections of the inspection report(s) attesting that the job is 100-percent complete and is correct per the job requirements found in this Scope of Work.

Detailed Job Requirements:

1. A new set of plans is required for each house. Plans are subject to changes and modifications. It is the responsibility of the Trade Contractor to have the new plans before beginning work. Plans should be picked up from the Site Superintendent. The Trade Contractor at no cost to The Company will correct any errors that occur from using an incorrect set of plans.

2. The Trade Contractor and the Site Superintendent must walk the job together and complete the pre-work section of the inspection report(s) before work may begin. The pre-work section of the inspection report(s) must be signed-off on by both parties.

3. All trim should be checked for damage prior to installation. Trim with small nicks, gouges, etc., that can be repaired so that damage is not noticeable should be used and repaired. Any defects that are visible from a distance of 6 feet in sunlight or normal house lighting after repairs are unacceptable and the trim piece should not be used.

_____ Company Rep's Initials
_____ Trade Contractor's Initials

4. Areas with vinyl shall receive shoe molding after the vinyl installation is complete.

5. Interior door units and exterior doors shall receive wood casing at the head and sides.

6. Fasteners shall be of the type specifically designed for trim. Fasteners shall be placed 18 inches on center. No fastener shall protrude from the trim and no fastener shall crack or splinter the wood. Any cracks or splinters that are small shall be repaired with wood filler, sanded, and smoothed. Any trim with damage that cannot be repaired to achieve an undamaged appearance under normal house lighting from a distance of 6 feet shall be replaced.

7. All rough areas on trim will be lightly sanded and smoothed.

8. All nicks, gouges, or blemishes shall be corrected by the use of wood filler, sanded, and smoothed.

9. All attic access or scuttle holes shall be trimmed as necessary.

10. If the house plan calls for cedar sheets in the master closet the sheets shall be installed according to plan.

11. All joints must meet flush with no visible gap

12. All base, chair rail, or crown material should be coped in corners or at miter joints

13. All crown and/or chair rails shall be installed level, straight, and plumb with no more than 1/4 inch deflection in 8 feet.

14. All base ending against doorjambs or other vertical areas shall be cut straight at a 90-degree angle.

15. Shoe molding shall be mitered at all corners (inside and out) to fit snugly.

16. All shoe molding shall be cut, when ending against doorjambs or openings with no jamb, at a small angle to achieve a smooth transition from the molding to the jamb or wall. The cuts are to be no more than 1/4 inch.

17. Door trim is to fit snugly to the carpet or vinyl. No visible gap between the vinyl and trim shall be allowed.

18. The Trade Contractor is responsible for the accurate height of all door openings over carpet and vinyl. If the door opening is inaccurate the Site Superintendent should be notified so that corrections can be made before the door is hung or trim installed, or the Trade Contractor can make the necessary adjustments. Doors that are not installed to the correct height are the responsibility of the trim Trade Contractor and not the framing Trade Contractor.

19. The base is to be installed level and plumb, and must fit snugly against all walls.

20. Interior doors are shipped prehung and prebored for locks. Interior doors should be installed level, plumb in both directions, and squarely in the opening with no more than 1/4 inch in 4 feet deviation in any direction.

Company Rep's Initials _____
Trade Contractor's Initials _____

21. Prehung doors should be checked for jamb squareness and straightness.

22. Reinstall manufacturer's braces to square units where necessary.

23. After leveling and plumbing the unit, shim as necessary to ensure correct installation.

24. All bifold doors shall hang squarely and open smoothly.

25. All doors are ordered with a 1-1/4-inch undercut in order to swing over carpeting. Additional undercutting will be performed as necessary by the Trade Contractor at no additional cost.

26. Any nicks, gouges, or damaged areas on door shall be repaired with the correct compound for the type of door used. Wood doors require the use of wood filler.

27. Exterior locks, interior locks, and other hardware shall be installed after the final painting is complete. All locks shall fit properly and lock smoothly. All doors are to operate easily and latch securely.

28. Hardware is furnished by The Company with construction locks in place on the exterior doors until after the homeowner's walk-through.

29. All bedroom and bathroom doors are to receive privacy locks.

30. Exterior doors will receive deadbolts keyed on both sides.

31. Stairs are prebuilt and installed by the stair Trade Contractor. The trim Trade Contractor is responsible for the installation of handrails, pickets, newel posts, and cover molding on the stair unit.

32. Handrails, pickets, and newel posts are to be installed securely, solidly, and have no movement.

33. Care should be taken to not damage the end of the stairs when attaching pickets and newel posts. Any damage must be repaired to present an undamaged appearance.

34. Cover molding must be installed securely to wall and to the skirt of the stair unit.

35. If the plans call for a raised fireplace hearth the Trade Contractor is responsible for constructing a box for the hearth. It is to be level, plumb, square, and of the correct size. The hearth box shall not have a deflection of more than 1/4 inch over the surface area of the box. All vertical areas are to be of the same height within 1/16 inch.

36. The Trade Contractor is expected to return to make any corrections to trim, doors, etc., after the homeowner's walk-through. All punch-out items to correct unacceptable work is the responsibility of the Trade Contractor and shall be done at no charge.

37. All construction debris shall be removed to the dumpster or designated trash area. The job will not be considered complete until the house and site are clean.

38. The house is to be broom-swept and clean before the job will be considered complete.

_____ Company Rep's Initials
_____ Trade Contractor's Initials

39. The Trade Contractor and Site Superintendent must walk the job together and perform a final inspection of the job. The final section of the inspection report(s) must be completed and signed-off on. The inspection report(s) must be attached to the office's copy of the purchase order and the Trade Contractor's invoice or payment will not be issued.

40. Any items found during the final inspection that need correction shall be corrected before payment will be made.

I _____ agent for _____

_____ have read and fully understand the above **Scope of Work** and I hereby agree to perform all work in accordance with the above.

Date: _____

Signed: Trade Contractor (or agent) _____

Date: _____

For The Company _____

INSPECTION REPORT

TRIM LABOR

Subdivision/ Lot # _____

Pre-Work Inspection: (to be completed prior to beginning work)

____ Plans have been picked up and signed for.

____ Purchase order has been received.

____ Material is on site and ready to be used.

____ House is clean, broom-swept, and free of debris.

____ Garage is clean.

____ Windows are intact and unbroken.

____ Stairs are installed.

____ Drywall is undamaged.

____ Vinyl is undamaged.

____ Ducts are covered.

____ Tubs are covered and undamaged.

____ Sinks, cabinets, and countertops are covered and undamaged.

____ Cabinets are set.

____ Doors are framed plumb and of correct size. Plan used to check carpet and vinyl areas.

____ Swing of doors is clearly marked on inside of each doorjamb.

____ Disappearing stairs are installed, if required.

____ Attic access is framed.

Date: _____   Site Superintendent _____

                             Trade Contractor _____

Final Inspection: (to be completed before Trade Contractor leaves job site)

____ Windows are intact and unbroken.

____ Drywall is undamaged.

____ Vinyl is undamaged.

____ Ducts are covered.

____ Tubs are covered and undamaged.

____ Sinks, cabinets, and countertops are covered and undamaged.

____ Base molding is installed securely to walls.

____ Base molding is undamaged and all nicks, gouges, etc., are repaired and sanded.

____ Base molding is level and straight, especially in tight and/or short runs.

____ Crown if required, is secure, level, and straight.

____ Chair rails, if required, are secure, level, and straight.

____ Nail holes are correct depth with no splintering or cracking.

____ Corners are cut correctly and fit snugly.

____ Doors are hung level, plumb, and shimmed.

____ Handrails are installed to the correct length, and are straight and secure.

____ Pickets and newels are installed correctly and securely with no movement.

____ Stair treads are undamaged.

____ Attic access areas are trimmed per plan.

____ Excess material has been removed to garage.

____ All debris has been removed to dumpster.

____ House is clean and broom-swept.

Date: _____   Site Superintendent _____

Trade Contractor _____

INSPECTION REPORT

TRIM LABOR (LOCK-OUT)

Subdivision/ Lot # _____

____	Doors open and close smoothly.
____	Doors latch correctly.
____	Privacy locks are installed on all bedrooms and baths.
____	Shoe molding is installed over all vinyl, securely and tightly to the base.
____	Shoe molding cuts at doors are cut 1/4 inch or less.
____	Corners of shoe molding fit snugly and are cut correctly.
____	Debris has been removed to dumpster.
____	House is broom-swept.

Date: _____   Site Superintendent _____

Trade Contractor _____

SCOPE OF WORK
Standards and Description of Work Performance

TERMITE TREATMENTS

The Company's **Terms and Conditions** are by reference a part of all **Scope of Work** requirements.

Construction Requirements:

Generally speaking, the work of Trade Contractors and their employees is expected to be performed in a good and workmanlike manner. Workmanlike quality is defined as workmanship which meets or betters those criteria indicated in applicable building codes, using materials and installation methods identified in the construction plans and this Scope of Work.

Code Requirements:

All jobs shall conform to those standards stipulated in the building code, mechanical code, plumbing code, and electrical code applicable in the local jurisdiction. All construction on The Company's jobsites shall meet or exceed NAHB Performance and Building Standards.

General Comments:

The Company considers our Trade Contractors to be experts at producing a high-quality job. But everyone on our construction team—staff, Trade Contractors, and suppliers   must recognize the importance of providing quality in both the product and service areas while on our jobsites and in the homes of our purchasers.

Since we work as a team, poor quality or service from any of us reflects unfavorably on all of us. An exceptional level of product quality and highly effective service can help us all to increase our business and grow.

The Company's definition of quality construction also requires that every job be completed correctly the first time. When this does not occur it costs both of us additional money, imposes on the purchaser, and hurts our reputations as quality builders. That is why, in situations where construction was not completed in a quality manner, prompt corrective action is required to remedy specific deficiencies.

In the following information the term Site Superintendent shall refer to any The Company representative with authority to perform the specified task. The term Trade Contractor shall mean the Trade Contractor's organization or any representative that is assigned the authority to perform the specified task.

General: All new homes built by The Company shall be pretreated against termites.

Material: All termiticides shall be of a type approved for the treatment of the soil to prevent termite infestation.

Company Rep's Initials _____
Trade Contractor's Initials _____

Installation: All treatments shall be done by experienced, trained personnel. Termiticide shall be applied to the soil before the polyethylene sheeting is placed. The termiticide should be of the strength suggested by the manufacturer and installation shall be per the manufacturer's treatment instructions.

A final treatment of termiticide shall be applied when the backfilling of the foundation is complete.

Other: Prior to the closing of the sale of the house the Trade Contractor shall be notified of the closing date and the name of the purchaser. The Trade Contractor must prepare an Infestation Certificate that is acceptable to FHA. A certificate acceptable to FHA is required on all The Company's houses, whether or not they are being sold FHA. Certificates shall be issued per Rule 620-6-03.

In any instance where the Trade Contractor finds an exception to the minimum treatment standard, the Trade Contractor shall forward to The Company a completed "Form II—Wood Destroying Organism Exception Report." The Company will in turn immediately correct the deficiency and will notify the Trade Contractor that the exception has been corrected. The Trade Contractor then shall reinspect the job to satisfy himself or herself that the exception has been corrected. The Company does not allow exceptions on Final Infestation Certificates.

Warranty: The Company believes that all work done and all materials installed in connection with one of our homes should be of high quality and that all Trade Contractors should stand behind the quality of their work and materials. Therefore we require all Trade Contractors to warrant the quality of their work for a period of one (1) year from date of closing of the house. Please refer to our printed Limited Warranty booklet for any specific items that are covered under Warranty as they apply to termite treatment.

The Trade Contractor shall have seven (7) days in which to correct any Warranty problem. If the problem is not corrected within seven (7) days then The Company shall correct the problem and will backcharge the Trade Contractor at the rate of $25.00 per hour, with a minimum charge of $100.00 plus the cost of any materials.

Inspection Reports: The Trade Contractor and the Site Superintendent shall walk the job together and complete each section of the inspection report(s). The Trade Contractor must correct any deficiency found during the inspection and the job must be 100-percent complete before payment will be made. The Trade Contractor and the Site Superintendent must sign-off on all sections of the inspection report(s) attesting that the job is 100-percent complete and is correct per the job requirements found in this Scope of Work.

Detailed Job Requirements:

1. Purchase orders will be mailed to the Trade Contractor from the office.

2. Termiticide is to be applied at the strength suggested by the manufacturer.

3. Termiticide shall be applied per the manufacturer's installation instructions.

4. Termiticide shall be applied to the soil before the polyethylene sheeting is applied.

_____ Company Rep's Initials
_____ Trade Contractor's Initials

5. The final termiticide treatment shall be applied when the backfilling of the foundation is complete.

6. Termite letters shall be issued that are acceptable to FHA with no exceptions listed.

I _____ agent for _____

_____ have read and fully understand the above **Scope of Work** and I hereby agree to perform all work in accordance with the above.

Date: _____   _____
                              Signed: Trade Contractor (or agent)

Date: _____   _____
                              For The Company

INSPECTION REPORT

TERMITE TREAMENTS

Subdivision/Lot # _____

Pre-Work Inspection: (to be completed prior to beginning work)

____ Purchase order has been received.

____ Lot is ready for first treatment.

____ Lot is clear of debris.

Date: _____ Site Superintendent _____

Trade Contractor _____

Final Inspection (first treatment): (to be completed before Trade Contractor leaves jobsite)

____ Soil has been treated.

Date: _____ Site Superintendent _____

Trade Contractor _____

Final Inspection (second treatment): (to be completed before Trade Contractor leaves jobsite)

____ Final treatment applied after backfill of slab.

Date: _____ Site Superintendent _____

Trade Contractor _____

SCOPE OF WORK
Standards and Description of Work Performance

VANITY TOPS

The Company's **Terms and Conditions** are by reference a part of all **Scope of Work** requirements.

Construction Requirements:

Generally speaking, the work of Trade Contractors and their employees is expected to be performed in a good and workmanlike manner. Workmanlike quality is defined as workmanship which meets or betters those criteria indicated in applicable building codes, using materials and installation methods identified in the construction plans and this Scope of Work.

Code Requirements:

All jobs shall conform to those standards stipulated in the building code, mechanical code, plumbing code and electrical code applicable in the local jurisdiction. All construction on The Company's jobsites shall meet or exceed NAHB Performance and Building Standards.

General Comments:

The Company considers our Trade Contractors to be experts at producing a high-quality job. But everyone on our construction team—staff, Trade Contractors, and suppliers—must recognize the importance of providing quality in both the product and service areas while on our jobsites and in the homes of our purchasers.

Since we work as a team, poor quality or service from any of us reflects unfavorably on all of us. An exceptional level of product quality and highly effective service can help us all to increase our business and grow.

The Company's definition of quality construction also requires that every job be completed correctly the first time. When this does not occur it costs both of us additional money, imposes on the purchaser and hurts our reputations as quality builders. That is why, in situations where construction was not completed in a quality manner, prompt corrective action is required to remedy specific deficiencies.

In the following information the term Site Superintendent shall refer to any The Company representative with authority to perform the specified task. The term Trade Contractor shall mean the Trade Contractor's organization or any representative that is assigned the authority to perform the specified task.

General: For most buyers, the look of finished carpentry, millwork, and cabinetry defines quality construction. The look, feel, and number of cabinets are major selling points for most homebuyers. The first impression created by the kitchen and baths must be a pleasant one. When looked at closely, these rooms must continue to be pleasant. Workmanship in these areas that is not high-quality turns-off homebuyers.

Company Rep's Initials _____
Trade Contractor's Initials _____

Care should be taken in the storage and handling of finish materials to avoid damage and soiling. Installed materials should be protected when necessary.

Materials: All cabinet materials, vanity tops, construction, type, etc., is determined by the model of the home, the location of the subdivision, and whether the cabinets are standard or upgrades.

Installation: All installations are to be performed by the Trade Contractor. All work is to be done by trained, experienced individuals. Vanity tops should be installed with sufficient care to avoid damage. All tops are to fit securely and snugly to the cabinets and walls. Tops are to be level, plumb, and square.

Care should be taken not to penetrate the vanity top's surface with fasteners. Vanity tops should be installed within 3/8 inch of level in 10 feet and 1/4 inch of level front to back.

Prior to the manufacture of vanity tops for the house, the Trade Contractor is responsible for field measuring all cabinet areas to ensure the correct fit of the tops.

Warranty: The Company believes that all work done and all materials installed in connection with one of our homes should be of high quality and that all Trade Contractors should stand behind the quality of their work and materials. Therefore we require all Trade Contractors to warrant the quality of their work for a period of one (1) year from date of closing of the house.

The trade contractor shall have seven (7) days in which to correct any Warranty problem. If the problem is not corrected within seven (7) days then The Company shall correct the problem and will backcharge the trade contractor at the rate of $25.00 per hour, with a minimum charge of $100.00 plus the cost of any materials.

Inspection Reports: The Trade Contractor and the Site Superintendent shall walk the job together and complete each section of the inspection report(s). The Trade Contractor must correct any deficiency found during the inspection and the job must be 100-percent complete before payment will be made. The Trade Contractor and the Site Superintendent must sign-off on all sections of the inspection report(s) attesting that the job is 100-percent complete and is correct per the job requirements found in this Scope of Work.

Detailed Job Requirements:

1. Plans are subject to changes and modifications. It is the responsibility of the Trade Contractor to verify with the Site Superintendent that the plans have not changed before beginning work. Any errors that occur from using an incorrect set of plans will be corrected by the Trade Contractor at no cost to The Company

2. Purchase orders will be mailed to the Trade Contractor from the office.

3. Color selection sheets detailing cabinet and countertop colors will be received from the Site Superintendent a minimum of three (3) weeks prior to installation.

4. The Trade Contractor and the Site Superintendent must walk the job together and complete the pre-work section of the inspection report(s) before work may begin. Both parties must sign-off on the pre-work section of the inspection report(s).

_____ Company Rep's Initials
_____ Trade Contractor's Initials

5. The Trade Contractor is responsible for field-measuring all cabinet areas to ensure proper sizing.

6. Vanity tops are to be built and installed to plan.

7. Vanity tops must match the color selection sheet specifications.

8. Vanity tops are to be installed per the manufacturer's installation instructions, but not with less quality than is stated above.

9. The Trade Contractor is responsible for the vanity tops being level, plumb, and securely attached. Vanity tops should be installed within 3/8 inch of level in 10 feet and 1/4 inch of level front to back.

10. After vanity tops are installed they should be protected with cardboard that is taped securely with a type of tape that will not damage the finish of the vanity top. Clear plastic sheeting taped to the tops is not acceptable as it routinely tears, splits, or otherwise is damaged, which in turn allows the vanity tops to be damaged.

11. All construction debris must be removed to the dumpster or to an area designated by the Site Superintendent. The job will not be considered complete and payment will not be issued until all trash and debris have been removed from the house and/or site.

12. The house is to be left clean and broom-swept.

13. The Trade Contractor and Site Superintendent must walk the job together and perform a final inspection of the job. The final section of the inspection report(s) must be completed and signed off on by both parties. The inspection report(s) must be attached to the office's copy of the purchase order and the Trade Contractor's invoice or payment will not be issued.

14. Correction of any items found on the final inspection report must be completed prior to payment being issued.

I _____ agent for _____

_____ have read and fully understand the above **Scope of Work** and I hereby agree to perform all work in accordance with the above.

Date: _____   _____
                            Signed: Trade Contractor (or agent)

Date: _____   _____
                            For The Company

INSPECTION REPORT

VANITY TOPS

Subdivision/Lot # _____

Pre-Work Inspection: (to be completed prior to beginning work)

____ Purchase order has been received.

____ Color selection sheet has been received.

____ Cabinet areas have been field-measured.

____ Windows are intact with none broken.

____ Tubs are covered and undamaged.

____ Temporary stairs, if required, temporary handrails, and safety bracing are installed.

____ House is broom-swept, clean, and free of debris.

Date: _____   Site Superintendent _____

Trade Contractor _____

Final Inspection: (to be completed before trade contractor leaves jobsite)

____ Windows are intact with none broken.

____ Temporary handrails, safety bracing, etc., are intact.

____ Vanity tops are installed per plan (correct number, size, color, etc.).

____ Tops are level and plumb.

____ Tops are secured correctly and tightly to wall.

____ Color(s) of vanity tops are correct.

____ Vanity tops are undamaged.

____ Vanity tops are protected with cardboard and taped securely.

____ House is clean, broom-swept, and debris has been removed.

Date: _____   Site Superintendent _____

Trade Contractor _____

<u>SCOPE OF WORK</u>
Standards and Description of Work Performance

WATERPROOFING

The Company's **Terms and Conditions** are by reference a part of all **Scope of Work** requirements.

Construction Requirements:

Generally speaking, the work of Trade Contractors and their employees is expected to be performed in a good and workmanlike manner. Workmanlike quality is defined as workmanship which meets or betters those criteria indicated in applicable building codes, using materials and installation methods identified in the construction plans and this Scope of Work.

Code Requirements:

All jobs shall conform to those standards stipulated in the building code, mechanical code, plumbing code and electrical code applicable in the local jurisdiction. All construction on The Company's jobsites shall meet or exceed NAHB Performance and Building Standards.

General Comments:

The Company considers our Trade Contractors to be experts at producing a high-quality job. But everyone on our construction team—staff, Trade Contractors, and suppliers—must recognize the importance of providing quality in both the product and service areas while on our jobsites and in the homes of our purchasers.

Since we work as a team, poor quality or service from any of us reflects unfavorably on all of us. An exceptional level of product quality and highly effective service can help us all to increase our business and grow.

The Company's definition of quality construction also requires that every job be completed correctly the first time. When this does not occur it costs both of us additional money, imposes on the purchaser, and hurts our reputations as quality builders. That is why, in situations where construction was not completed in a quality manner, prompt corrective action is required to remedy specific deficiencies.

In the following information the term Site Superintendent shall refer to any The Company representative with authority to perform the specified task. The term Trade Contractor shall mean the Trade Contractor's organization or any representative that is assigned the authority to perform the task.

General: The Trade Contractor is responsible for ensuring that all drain pipe and waterproofing materials are installed according to the manufacturer's specifications and instructions. The installation of the waterproofing material shall be such that no leakage occurs. The installation of drain pipe shall be such that no water is allowed against the foundation. All materials are to be supplied and installed by the Trade Contractor.

Company Rep's Initials _____
Trade Contractor's Initials _____

Foundation drains shall be provided around all concrete or masonry foundations enclosing basements (finished or unfinished) or any portion of the house that is located below the grade line.

Material: Foundation drains shall be 4-inch plastic drain pipe designed for residential foundation drains. Drainage pipe designed for septic tanks, etc., shall not be used. Waterproofing material shall be approved for residential use and shall be a spray-on type. Brush-on type waterproofing material shall not be used.

Installation: Foundation drains shall be installed at or below the area to be protected and shall discharge by gravity. Drain pipe must be placed on a minimum of 2 inches of washed gravel and covered with a minimum of 4 inches of washed gravel. The drain pipe is to be located at the side of the footing, not where the footing and foundation wall meet. Foundation walls should be clean. There should be no excess mortar, dirt, or any foreign materials on the foundation walls prior to applying waterproofing material. Waterproofing shall be installed per the manufacturer's installation instructions and in such a manner that the material is of a uniform thickness with no thin spots, runs, or lumps.

All miscellaneous debris, vegetation debris, and wood debris must be removed from the foundation trench. Any of these items left in the foundation trench will invalidate the termite/pest treatment.

Warranty: The Company believes that all work done and all materials installed in connection with one of our homes should be of high quality and that all Trade Contractors should stand behind the quality of their work and materials. Therefore we require all Trade Contractors to warrant the quality of their work for a period of one (1) year from date of closing of the house. Please refer to our printed Limited Warranty booklet for specific items that are covered under Warranty as they apply to waterproofing.

The Trade Contractor shall have seven(7) days in which to correct any Warranty problem. If the problem is not corrected within seven (7) days then The Company shall correct the problem and will backcharge the Trade Contractor at the rate of $25.00 per hour, with a minimum charge of $100.00 plus the cost of any materials.

Inspection Reports: The Trade Contractor and the Site Superintendent shall walk the job together and complete each section of the inspection report(s). The Trade Contractor must correct any deficiency found during the inspection and the job must be 100-percent complete before payment will be made. The Trade Contractor and the Site Superintendent must sign-off on all sections of the inspection report(s) attesting that the job is 100-percent complete and is correct per the job requirements found in this Scope of Work.

Detailed Job Procedures:

1. Plans are subject to changes and modifications. It is the responsibility of the Trade Contractor to verify with the Site Superintendent that the plans have not changed before beginning work. Any errors that occur from using an incorrect set of plans will be corrected by the Trade Contractor at no cost to The Company.

2. The Trade Contractor must pick up a copy of purchase order.

3. The Trade Contractor and the Site Superintendent must walk the job together and complete the pre-work section of the inspection report(s) before work may begin. Both parties must sign-off on the pre-work section of the inspection report(s).

_____ Company Rep's Initials
_____ Trade Contractor's Initials

4. Foundation drains shall be provided around all concrete or masonry foundations enclosing basements (finished or unfinished) or any portion of the house that is located below the grade line.

5. Foundation drains shall be 4-inch plastic drain pipe of a type designed for residential use. Drain pipe such as septic tank drain lines shall not be used.

6. Foundation drain pipe shall be installed below the area to be protected and shall discharge by gravity.

7. Drain pipe must be placed on a minimum of 2 inches of washed gravel and covered by a minimum of 4 inches of washed gravel.

8. All drain pipe is to extend a minimum of 2 feet beyond the outside edge of the footing and 6 inches above the top of the footing.

9. Pipe shall terminate to daylight. The end of drain pipe should be flagged to prevent other Trade Contractors from covering the drain pipe or filling in the open end of the line.

10. Foundation walls must be clean and free of excess mortar runs or other matter prior to the installation of waterproofing material.

11. All masonry units or poured concrete walls shall be fully coated with a sprayed-on asphalt waterproofing to the thickness required and using the installation procedures specified by the manufacturer. All coatings must be of a uniform thickness with no thin or missed spots.

12. The Trade Contractor is responsible for cleaning up of all residual materials before the job will be accepted as complete.

13. The Trade Contractor and Site Superintendent must walk the job together and perform a final inspection of the job. The final section of the inspection report(s) must be completed and signed-off on by both parties. The inspection report(s) must be attached to the office's copy of the purchase order and the Trade Contractor's invoice or payment will not be issued.

14. The Trade Contractor and the Site Superintendent together must field-measure the foundation footage and both must sign the Trade Contractor's invoice verifying the accuracy of the footage before payment will be made. Price is based on a per-lineal foot of the foundation.

15. Any items that require correction during the final inspection must be corrected before the job will be considered complete.

I _____ agent for _____

_____ have read and fully understand the above **Scope of Work** and I hereby agree to perform all work in accordance with the above.

Date: _____   _____
                           Signed: Trade Contractor (or agent)

Date: _____   _____
                           For The Company

INSPECTION REPORT

WATERPROOFING

Subdivision/Lot # _____

Pre-Work Inspection: (to be completed prior to beginning work)

____ Purchase order has been picked up.

____ Work area is clean and free of debris.

____ Foundation trench is free and clear of all debris.

____ All formboards have been removed.

____ Sides of foundation trench are sloped away from foundation to prevent cave-in.

____ No honeycombs are on poured walls.

____ No cracks are on masonry foundations or poured walls.

____ Foundation is clean and free of excess mortar, dirt, mud, etc.

Date: _____   Site Superintendent _____

                                                    Trade Contractor _____

Final Inspection: (to be completed before Trade Contractor leaves jobsite)

____ Waterproofing fully covers all foundation area from footing to top of grade.

____ Waterproofing is of consistent thickness with no thin or missed spots.

____ Stone is 2 inches below drain pipe and 4 inches above drain pipe.

____ Drain pipe extends a minimum of 2 feet beyond the outside edge of the footing and 6 inches above the top of the footing.

____ Work area is clear of all building debris.

____ Trade Contractor and Site Superintendent together have verified lineal footage of foundation.

____ All drain pipe are flagged to prevent damage by other Trade Contractors.

Date: _____   Site Superintendent _____

                                                    Trade Contractor _____

SCOPE OF WORK
Standards and Description of Work Performance

WINDOW AND DOOR INSTALLATION

The Company's **Terms and Conditions** are by reference a part of all **Scope of Work** requirements.

Construction Requirements:

Generally speaking, the work of Trade Contractors and their employees is expected to be performed in a good and workmanlike manner. Workmanlike quality is defined as workmanship which meets or betters those criteria indicated in applicable building codes, using materials and installation methods identified in the construction plans and this Scope of Work.

Code Requirements:

All jobs shall conform to those standards stipulated in the building code, mechanical code, plumbing code and electrical code applicable in the local jurisdiction. All construction on The Company's jobsites shall meet or exceed NAHB Performance and Building Standards.

General Comments:

The Company considers our Trade Contractors to be experts at producing a high-quality job. But everyone on our construction team—staff, Trade Contractors, and suppliers   must recognize the importance of providing quality in both the product and service areas while on our jobsites and in the homes of our purchasers.

Since we work as a team, poor quality or service from any of us reflects unfavorably on all of us. An exceptional level of product quality and highly effective service can help us all to increase our business and grow.

The Company's definition of quality construction also requires that every job be completed correctly the first time. When this does not occur it costs both of us additional money, imposes on the purchaser, and hurts our reputations as quality builders. That is why, in situations where construction was not completed in a quality manner, prompt corrective action is required to remedy specific deficiencies.

In the following information the term Site Superintendent shall refer to any The Company representative with authority to perform the specified task. The term Trade Contractor shall mean the Trade Contractor's organization or any representative that is assigned the authority to perform the specified task.

General: All windows and doors shall be per the specifications furnished the manufacturer by The Company and shall be installed per plan.

The furnishing and installing of windows and doors shall be considered a turnkey job and the Trade Contractor is responsible for furnishing the correctly sized, type, and grade of windows and doors. All installations shall be the responsibility of the Trade Contractor.

Company Rep's Initials _____
Trade Contractor's Initials _____

Inspection Reports: The Trade Contractor and the Site Superintendent shall walk the job together and complete each section of the inspection report(s). The Trade Contractor must correct any deficiency found during the inspection and the job must be 100-percent complete before payment will be made. The Trade Contractor and the Site Superintendent must sign-off on all sections of the inspection report(s) attesting that the job is 100-percent complete and is correct per the job requirements found in this Scope of Work.

Detailed Job Requirements:

1. A new set of plans is required for each house. Plans are subject to changes and modifications. It is the responsibility of the Trade Contractor to have the new plans before beginning work. Plans should be picked up at the job trailer from the Site Superintendent. Any errors that occur from using an incorrect set of plans will be corrected by the Trade Contractor at no cost to The Company.

2. Purchase orders shall be mailed to the Trade Contractor from the office and shall be sent in ample time for the Trade Contractor to have the correct windows and doors on hand.

3. The Trade Contractor and the Site Superintendent must walk the job together and complete the pre-work section of the inspection report(s) before work may begin. Both parties must sign-off on the pre-work section of the inspection report(s).

4. Windows and doors shall be installed per plan and per the manufacturer's instructions.

5. If any framed opening is not correctly sized, or is out of square or not level, and the Trade Contractor cannot easily correct the problem, the Site Superintendent should be immediately notified.

6. All windows and doors shall be installed so that they are centered in the framed opening and level within 1/16 of an inch.

7. All windows and doors shall operate smoothly and correctly.

8. Exterior door warpage of more than 1/4 inch measured diagonally from corner to corner is unacceptable.

9. Weatherstripping, insulation, and waterproofing shall be installed per the manufacturer's requirements. Air leakage is unacceptable

10. Screens shall be installed on windows and any doors requiring screens prior to the homeowner's walk-through.

11. The Trade Contractor and Site Superintendent must walk the job together and perform a final inspection of the job. The final section of the inspection report(s) must be completed and signed-off on by both parties. The inspection report(s) must be attached to the office's copy of the purchase order and the Trade Contractor's invoice or payment will not be issued.

12. Any items found during the final inspection that need correction shall be corrected before payment will be made.

_____ Company Rep's Initials
_____ Trade Contractor's Initials

I _____ agent for _____

_____ have read and fully understand the above **Scope of Work** and I hereby agree to perform all work in accordance with the above.

Date: _____   _____

                            Signed: Trade Contractor (or agent)

Date: _____   _____

                            For The Company

INSPECTION REPORT

WINDOW and DOOR INSTALLATION

Subdivision/Lot # _____

Pre-Work Inspection: (to be completed prior to beginning work)

____ Plans have been received.

____ Purchase order has been received.

____ House is clean and free of debris.

____ Framing is complete.

____ Window and door openings have been checked and dimensions are correct.

____ Electricity is available.

Date: _____   Site Superintendent _____

Trade Contractor _____

Final Inspection: (to be completed before Trade Contractor leaves jobsite)

____ All windows and doors are set.

____ There are no broken windows.

____ All windows operate and lock correctly.

____ Doors operate correctly.

____ Door swings are correct.

____ Weatherstripping is installed.

____ There is no air leakage around windows or doors.

____ All windows are plumb, level, and square.

____ All doors are plumb, level, and square.

____ House is free and clear of debris.

____ House is broom-swept.

Date: _____   Site Superintendent _____

Trade Contractor _____

Appendix B

Associated Documents

The scopes of work program includes two sets of terms and conditions and a general terms and conditions that are referenced in the scopes and must be included in the documents that your Trade Contractors sign. The following terms and conditions and purchase orders can be found on the CD. I suggest that you print these documents off of the CD for review by your technical and management staff, and by your company's attorney. See pages 231-232 for the document and its corresponding file name on the CD, as well as information on using the CD.

1. **Terms and Conditions**: This document covers what paperwork the builder requires of all trade contractors before they are allowed to begin work. Items covered include insurance, proof of IRS business number or social security number, safety program, and the builder's policy for a drug-free workplace.

2. **General Terms and Conditions:** This document covers the rules and regulations for the builder's jobsite and jobs. All people, trade contractors or suppliers, are expected to abide by these rules and regulations. The rules and regulations also detail what these same people, trade contractors and suppliers, may expect from the builder. Items covered in the sample include construction requirements, punch-out lists, homeowners' walk-throughs, safety, insurance requirements, purchase orders, cleanliness, warranty, completeness of work, drug-free workplace, pricing and change orders, and waste of materials. This set of terms and conditions should be printed on the back of purchase orders.

   There is a signature form included with the two terms and conditions that must be signed by both the trade contractor and a company representative. Both of these terms-and-conditions documents should be reviewed by your attorney to ensure they are enforceable in your state.

3. **Purchase Orders:** Many builders do not use preprinted purchase orders; some do not use purchase orders at all. If you would like to begin using preprinted purchase orders, I have included a sample purchase order for suppliers and another for trade contractors. The purchase order for trade contractors is set up so that the trade contractor can use it for their invoice. (We all get rather tired of seeing invoices written on scraps of paper.) The purchase order for trade contractors also includes a lien waiver on the bottom. Ask your attorney to verify whether or not the wording is legal in your state and, if necessary, to make the required modifications.

   You do not have to have a computer system to use preprinted purchase orders. If you normally write purchase orders by hand using a purchase order book, exchange the purchase order pad for preprinted purchase orders and continue to write your purchase orders by hand. Make a photocopy to give to the trade contractor or supplier.

Terms and Conditions

There are certain **Terms and Conditions** that **The Company** requires of all **Trade Contractors**. Please read the following carefully before you sign. A copy of these Terms and Conditions should be given to all of your onsite supervisors. The term "you" and /or the term "the Trade Contractor" as used in these General Terms and Conditions means not only you as an individual but also any and all employees or trade contractors that you use in the completion of your Scope of Work for The Company. The term "we" as used below shall mean The Company.

GENERAL INFORMATION

- **Insurance:** Prior to you being allowed to begin work an original Certificate of Insurance, from your insurance agent, reflecting current Worker's Compensation and General Liability Insurance must be in our office.

    It is your responsibility to notify us immediately if your insurance is canceled for any reason. No Trade Contractor is allowed to work without current insurance. Should you fail to notify us and we are notified of your uninsured status by someone other than you, you will be removed from our list of approved Trade Contractors and all money due you for any work will be withheld, in full, until you have current insurance and we have received a new original Certificate of Insurance.

- **Proof of Name and IRS ID Number:** The IRS requires that we have on file in our office proof that the name and Federal ID number (or Social Security number) that you are working under is the same as is on file with the IRS. We require a copy of a document from the IRS showing your name and ID number (Federal ID for businesses or Social Security number for individuals). You also will be asked to complete a Form W-9. Without this documentation the IRS requires that we withhold 20 percent of all money due you which must then be forwarded to the IRS.

    The "insured name" on your insurance certificate must match the name on the documents above and the W-9. All checks will be made payable to the name on the above documents. No insurance certificate will be accepted as valid that reflects a different name from the one on the above described documents.

- **Safety:** We have a full Safety and Haz-Com program. You, as a trade contractor on our jobsite(s), are required to have your own Safety and Haz-Com programs. We are required to have a copy of your programs on file, in our office, and also in the job shack at any of our jobsite(s) you at which you work. For your own protection you should have a copy of these programs and the necessary Haz-Com data sheets in the truck of each of your onsite supervisors.

- **Drug-Fee Workplace:** We are a drug-free workplace. We do not allow the use of controlled substances (drugs) or alcohol on any of our jobsites. Should any of your employees be found to be in possession of either drugs or alcohol on our jobsites you will be requested to leave the jobsite and will not be allowed to return to work until you have corrected the problem.

_____ Company Rep's Initials
_____ Trade Contractor's Initials

General Terms and Conditions

This section of the General Terms and Conditions will be found on the back of all purchase orders. You will be required to sign the purchase order before payment will be issued. Please be sure you understand these Terms and Conditions.

Construction Requirements: The work of all Trade Contractors, their employees, and/or trades is expected to be performed in a good and workmanlike manner. Workmanlike quality is defined as workmanship which meets or betters those criteria indicated in the building codes, using materials and installation methods identified in the construction plans, The Company's Scope of Work and defined by industry standards for each trade.

Inspection Reports: The Trade Contractor and a Company representative shall walk the job together and complete each section of the inspection report(s). The Trade Contractor must correct any deficiency found during the inspection and the job must be 100-percent complete before payment will be made. The Trade Contractor and a Company representative must sign-off on all sections of the inspection report(s) attesting that the job is correct and complete.

Punch-List: The site superintendent shall inspect the work of each Trade Contractor and will issue a punch-list of all items requiring correction as soon as the Trade Contractor has completed his or her work. The Trade Contractor is expected to immediately complete their punch-list so that he or she does not slow down overall construction of the home and/or the scheduling of the next trade contractor. Should the Trade Contractor not return to complete his or her punch-list within a reasonable time then the site superintendent may hire someone to complete the punch-list and the Trade Contractor will be backcharged for this work. The job will not be considered to be complete and no payment shall be made until all punch-list items are completed and approved by the site superintendent.

Completeness of Work: Invoices must be turned into the site superintendent by no later than Thursday at 5:00 p.m. Invoices will be approved and turned into the office on Friday. Payment will be ready to be picked up the following Friday. The Trade Contractor or their representative must sign his or her purchase order before payment will be issued. The purchase order may be used in place of an invoice to request payment.

Homeowner's Walk-Through List: Any items found on the homeowner's walk-through that require correction must be completed immediately upon notification by the site superintendent. Time is of the essence for completing corrections on homeowner's walk-throughs. If any Trade Contractor does not return to correct his or her items listed on the walk-through list then someone else will be hired to make the corrections and the Trade Contractor will be backcharged.

Safety: The Trade Contractor acknowledges that he or she has completed the Safety Training program as required by The Company and that he or she has his or her own Safety and Haz-Com program for their employees and/or trade contractors. Trade Contractor agrees to comply with OSHA and/or any other governmental agency's safety rules and regulations. Should any citations, fines, and/or penalties, etc., be incurred by The Company due to the negligence of the Trade Contractor, the Trade Contractor agrees to indemnify The Company for any and all penalties, fines, etc., incurred.

Insurance: Trade Contractor acknowledges that a requirement of working for The Company is for the Trade Contractor to have current Worker's Compensation and General Liability Insurance at all times. The Trade Contractor agrees to indemnify The Company and to be responsible for any claims, expenses, or litigation arising from any claim made against The Company due to any injury of the Trade Contractor's employee or trade contractor for any worker's compensation claim. The Trade Contractor also agrees to indemnify The Company and to be responsible for any claims, expenses, or litigation arising from any claim made against The Company due to the workmanship, equipment, or materials supplied by the Trade Contractor.

Purchase Order Number: No invoice shall be paid that does not include the purchase order number for the job. The preprinted purchase order issued by The Company may be used as an invoice.

Cleanliness: The Trade Contractor is responsible for leaving the work area clean and free of debris. If it is necessary for The Company to remove debris left by the Trade Contractor, the Trade Contractor will be assessed a clean-up fee of $100.00. The site superintendent or other The Company representative will designate an area for all building debris and trash to be placed. Such area may be a dumpster or a designated trash pile on the lot.

Trash, such as lunch or snack trash, is not to be thrown on the floor of the house or on the jobsite. All such trash must be placed in the trash basket/can

Port-a-johns are provided on all job sites. Any person found using sinks, tubs, commodes that are not hooked up, ductwork, closets, etc., as a toilet will be fined $100.00 and will not be allowed back on any of The Company's jobsites. It is the responsibility of the Trade Contractor to impress on his or her employees and trade contractors that this offensive habit of using any area as a toilet facility will not be tolerated.

Warranty: All work is to be guaranteed for one (1) year from date of closing of the house. Certain items must be guaranteed for (2) years. These items are detailed in The Company's printed Limited Warranty booklet. The Trade Contractor acknowledges that he or she received a copy of The Company's printed Limited Warranty booklet and that he or she agrees to abide by the warranty coverage requirements and time period printed in this document as they pertain to his or her trade.

Company Rep's Initials _____
Trade Contractor's Initials _____

Drug-Free Workplace: The Company is a drug-free workplace. The use of any controlled substances (drugs) or alcohol on any of The Company's jobsite(s) is not permitted. Should the Trade Contractor, his or her employees and/or trade contractors be found to be in possession of either drugs or alcohol on the jobsite(s) the Trade Contractor will be requested to leave the jobsite and will not be allowed to return to work until the problem is corrected.

Pricing/Change Orders: All work is quoted and priced **per model**. Payment will be made per the price listed on the purchase order. Any change order will be priced per change order. No additional work will be considered, allowed, or paid other than that priced on the purchase order. Should the Trade Contractor be requested to perform any additional work the Trade Contractor must request a hand purchase order from the site superintendent.

Damage and/or Wastefulness of Materials: Damage to materials and installed items such as carpet, vinyl, fixtures, etc., caused by negligence on the part of the Trade Contractor, his or her employees and/or trade contractors will result in backcharges for the amount necessary to replace or repair the item. Wastefulness of materials by the Trade Contractor will result in the cost of that material being deducted from payment due the Trade Contractor.

General Terms and Conditions

Trade Contractor's Company Name: _____

Mailing Address: _____

Phone: Day _____   Night _____   Beeper _____

I _____ the _____ for the above named Trade Contractor agree that I have read and fully understand the **General Terms and Conditions** attached to this page and made a part of any agreement I have with The Company. I agree that I have received a copy of these **General Terms and Conditions** and I agree to abide by these **General Terms and Conditions**. I understand that I am responsible for any employee or trade contractor that I bring onto the jobsite and that they must abide by these **General Terms and Conditions**.

Signed for/by Trade Contractor _____   Date _____

Signed for/by The Company _____   Date _____

Witness: _____

PURCHASE
ORDER

TO:

Ship To:

P.O. No.	P.O. Date	Model No	Description	Date Req.

IMPORTANT: This Purchase Order is subject to the Terms and Conditions signed by the Subcontractor as a condition for performing any work for The Company and as printed on the reverse side hereof for this job.

In-house #	Cost Code	Description of Job	Unit	Quantity	Unit Price	Total

I the undersigned Trade Contractor certify that all materials, labor, and services furnished by me or to me for the above referenced job have been fully paid for and that the premises on the above job cannot be made subject to any valid lien or claim by anyone who furnished material, labor, or services to the Trade Contractor for use in said job and the Trade Contractor hereby releases The Company from any further liability in connection with all materials, labor, and services. This release is given in order to induce payment in the amount stated above and upon receipt of said payment to the Trade Contractor this release becomes in full force and effect.

I, the Trade Contractor, certify that I have performed the work and installed the materials and/or equipment specified in the Scope of Work and bid/quote for the above referenced job identified above, and before requesting payment I certify that I have inspected the completed work and hereby warrant, as a condition of receiving payment, that the materials and workmanship meet or exceed the Scope of Work conditions and specifications.

_____
Site Superintendent

(By signing above the site superintendent certifies that he or she has inspected the above job and it is complete including all punch-list items and payment are hereby OK'ed).

Printed Name: _____

Date: _____

_____
Trade Contractor

By:_____

Printed Name: _____

Terms and Conditions

Construction Requirements: The work of all Trade Contractors, their employees, and/or trades is expected to be performed in a good and workmanlike manner. Workmanlike quality is defined as workmanship which meets or betters those criteria indicated in the building codes, using materials and installation methods identified in the construction plans, The Company's Scope of Work and defined by industry standards for each trade.

Inspection Reports: The Trade Contractor and a Company representative shall walk the job together and complete each section of the inspection report(s). The Trade Contractor must correct any deficiency found during the inspection and the job must be 100-percent complete before payment will be made. The Trade Contractor and a Company representative must sign-off on all sections of the inspection report(s) attesting that the job is correct and complete.

Punch-List: The site superintendent shall inspect the work of each Trade Contractor and will issue a punch-list of all items requiring correction as soon as the Trade Contractor has completed his or her work. The Trade Contractor is expected to immediately complete their punch-list so that he or she does not slow down overall construction of the home and/or the scheduling of the next trade contractor. Should the Trade Contractor not return to complete his or her punch-list within a reasonable time then the site superintendent may hire someone to complete the punch-list and the Trade Contractor will be backcharged for this work. The job will not be considered to be complete and no payment shall be made until all punch-list items are completed and approved by the site superintendent.

Completeness of Work: Invoices must be turned into the site superintendent by no later than Thursday at 5:00 p.m. Invoices will be approved and turned into the office on Friday. Payment will be ready to be picked up the following Friday. The Trade Contractor or their representative must sign his or her purchase order before payment will be issued. The purchase order may be used in place of an invoice to request payment.

Homeowner's Walk-Through List: Any items found on the homeowner's walk-through that require correction must be completed immediately upon notification by the site superintendent. Time is of the essence for completing corrections on homeowner's walk-throughs. If any Trade Contractor does not return to correct his or her items listed on the walk-through list then someone else will be hired to make the corrections and the Trade Contractor will be backcharged.

Safety: The Trade Contractor acknowledges that he or she has completed the Safety Training program as required by The Company and that he or she has his or her own Safety and Haz-Com program for their employees and/or trade contractors. Trade Contractor agrees to comply with OSHA and/or any other governmental agency's safety rules and regulations. Should any citations, fines, and/or penalties, etc., be incurred by The Company due to the negligence of the Trade Contractor, the Trade Contractor agrees to indemnify The Company for any and all penalties, fines, etc., incurred.

Insurance: Trade Contractor acknowledges that a requirement of working for The Company is for the Trade Contractor to have current Worker's Compensation and General Liability Insurance at all times. The Trade Contractor agrees to indemnify The Company and to be responsible for any claims, expenses, or litigation arising from any claim made against The Company due to any injury of the Trade Contractor's employee or trade contractor for any worker's compensation claim. The Trade Contractor also agrees to indemnify The Company and to be responsible for any claims, expenses, or litigation arising from any claim made against The Company due to the workmanship, equipment, or materials supplied by the Trade Contractor.

Purchase Order Number: No invoice shall be paid that does not include the purchase order number for the job. The preprinted purchase order issued by The Company may be used as an invoice.

Cleanliness: The Trade Contractor is responsible for leaving the work area clean and free of debris. If it is necessary for The Company to remove debris left by the Trade Contractor, the Trade Contractor will be assessed a clean-up fee of $100.00. The site superintendent or other The Company representative will designate an area for all building debris and trash to be placed. Such area may be a dumpster or a designated trash pile on the lot.

Trash, such as lunch or snack trash, is not to be thrown on the floor of the house or on the jobsite. All such trash must be placed in the trash basket/can

Port-a-johns are provided on all job sites. Any person found using sinks, tubs, commodes that are not hooked up, ductwork, closets, etc., as a toilet will be fined $100.00 and will not be allowed back on any of The Company's jobsites. It is the responsibility of the Trade Contractor to impress on his or her employees and trade contractors that this offensive habit of using any area as a toilet facility will not be tolerated.

Warranty: All work is to be guaranteed for one (1) year from date of closing of the house. Certain items must be guaranteed for (2) years. These items are detailed in The Company's printed Limited Warranty booklet. The Trade Contractor acknowledges that he or she received a copy of The Company's printed Limited Warranty booklet and that he or she agrees to abide by the warranty coverage requirements and time period printed in this document as they pertain to his or her trade.

Drug-Free Workplace: The Company is a drug-free workplace. The use of any controlled substances (drugs) or alcohol on any of The Company's jobsite(s) is not permitted. Should the Trade Contractor, his or her employees and/or trade contractors be found to be in possession of either drugs or alcohol on the jobsite(s) the Trade Contractor will be requested to leave the jobsite and will not be allowed to return to work until the problem is corrected.

Pricing/Change Orders: All work is quoted and priced **per model.** Payment will be made per the price listed on the purchase order. Any change order will be priced per change order. No additional work will be considered, allowed, or paid other than that priced on the purchase order. Should the Trade Contractor be requested to perform any additional work the Trade Contractor must request a hand purchase order from the site superintendent.

Damage and/or Wastefulness of Materials: Damage to materials and installed items such as carpet, vinyl, fixtures, etc., caused by negligence on the part of the Trade Contractor, his or her employees and/or trade contractors will result in backcharges for the amount necessary to replace or repair the item. Wastefulness of materials by the Trade Contractor will result in the cost of that material being deducted from payment due the Trade Contractor.

PURCHASE
ORDER

TO: **Ship To:**

P.O. No.	P.O. Date	Model No	Description	Date Req.

In-house #	Cost Code	Description of Materials	Unit	Quantity	Unit Price	Total

The above shipment was received in good condition. All items listed were received except for those items marked above.

Date:_____   Site Superintendent: _____

Print Name: _____

Terms and Conditions

Construction Requirements: The work of all Trade Contractors, their employees, and/or trades is expected to be performed in a good and workmanlike manner. Workmanlike quality is defined as workmanship which meets or betters those criteria indicated in the building codes, using materials and installation methods identified in the construction plans, The Company's Scope of Work and defined by industry standards for each trade.

Inspection Reports: The Trade Contractor and a Company representative shall walk the job together and complete each section of the inspection report(s). The Trade Contractor must correct any deficiency found during the inspection and the job must be 100-percent complete before payment will be made. The Trade Contractor and a Company representative must sign-off on all sections of the inspection report(s) attesting that the job is correct and complete.

Punch-List: The site superintendent shall inspect the work of each Trade Contractor and will issue a punch-list of all items requiring correction as soon as the Trade Contractor has completed his or her work. The Trade Contractor is expected to immediately complete their punch-list so that he or she does not slow down overall construction of the home and/or the scheduling of the next trade contractor. Should the Trade Contractor not return to complete his or her punch-list within a reasonable time then the site superintendent may hire someone to complete the punch-list and the Trade Contractor will be backcharged for this work. The job will not be considered to be complete and no payment shall be made until all punch-list items are completed and approved by the site superintendent.

Completeness of Work: Invoices must be turned into the site superintendent by no later than Thursday at 5:00 p.m. Invoices will be approved and turned into the office on Friday. Payment will be ready to be picked up the following Friday. The Trade Contractor or their representative must sign his or her purchase order before payment will be issued. The purchase order may be used in place of an invoice to request payment.

Homeowner's Walk-Through List: Any items found on the homeowner's walk-through that require correction must be completed immediately upon notification by the site superintendent. Time is of the essence for completing corrections on homeowner's walk-throughs. If any Trade Contractor does not return to correct his or her items listed on the walk-through list then someone else will be hired to make the corrections and the Trade Contractor will be backcharged.

Safety: The Trade Contractor acknowledges that he or she has completed the Safety Training program as required by The Company and that he or she has his or her own Safety and Haz-Com program for their employees and/or trade contractors. Trade Contractor agrees to comply with OSHA and/or any other governmental agency's safety rules and regulations. Should any citations, fines, and/or penalties, etc., be incurred by The Company due to the negligence of the Trade Contractor, the Trade Contractor agrees to indemnify The Company for any and all penalties, fines, etc., incurred.

Insurance: Trade Contractor acknowledges that a requirement of working for The Company is for the Trade Contractor to have current Worker's Compensation and General Liability Insurance at all times. The Trade Contractor agrees to indemnify The Company and to be responsible for any claims, expenses, or litigation arising from any claim made against The Company due to any injury of the Trade

Contractor's employee or trade contractor for any worker's compensation claim. The Trade Contractor also agrees to indemnify The Company and to be responsible for any claims, expenses, or litigation arising from any claim made against The Company due to the workmanship, equipment, or materials supplied by the Trade Contractor.

Purchase Order Number: No invoice shall be paid that does not include the purchase order number for the job. The preprinted purchase order issued by The Company may be used as an invoice.

Cleanliness: The Trade Contractor is responsible for leaving the work area clean and free of debris. If it is necessary for The Company to remove debris left by the Trade Contractor, the Trade Contractor will be assessed a clean-up fee of $100.00. The site superintendent or other The Company representative will designate an area for all building debris and trash to be placed. Such area may be a dumpster or a designated trash pile on the lot.

Trash, such as lunch or snack trash, is not to be thrown on the floor of the house or on the jobsite. All such trash must be placed in the trash basket/can

Port-a-johns are provided on all job sites. Any person found using sinks, tubs, commodes that are not hooked up, ductwork, closets, etc., as a toilet will be fined $100.00 and will not be allowed back on any of The Company's jobsites. It is the responsibility of the Trade Contractor to impress on his or her employees and trade contractors that this offensive habit of using any area as a toilet facility will not be tolerated.

Warranty: All work is to be guaranteed for one (1) year from date of closing of the house. Certain items must be guaranteed for (2) years. These items are detailed in The Company's printed Limited Warranty booklet. The Trade Contractor acknowledges that he or she received a copy of The Company's printed Limited Warranty booklet and that he or she agrees to abide by the warranty coverage requirements and time period printed in this document as they pertain to his or her trade.

Drug-Free Workplace: The Company is a drug-free workplace. The use of any controlled substances (drugs) or alcohol on any of The Company's jobsite(s) is not permitted. Should the Trade Contractor, his or her employees and/or trade contractors be found to be in possession of either drugs or alcohol on the jobsite(s) the Trade Contractor will be requested to leave the jobsite and will not be allowed to return to work until the problem is corrected.

Pricing/Change Orders: All work is quoted and priced **per model**. Payment will be made per the price listed on the purchase order. Any change order will be priced per change order. No additional work will be considered, allowed, or paid other than that priced on the purchase order. Should the Trade Contractor be requested to perform any additional work the Trade Contractor must request a hand purchase order from the site superintendent.

Damage and/or Wastefulness of Materials: Damage to materials and installed items such as carpet, vinyl, fixtures, etc., caused by negligence on the part of the Trade Contractor, his or her employees and/or trade contractors will result in backcharges for the amount necessary to replace or repair the item. Wastefulness of materials by the Trade Contractor will result in the cost of that material being deducted from payment due the Trade Contractor.

How to Use *Scopes of Work* on CD

This CD contains the scopes of work and inspection reports formatted in Microsoft Word 97/Windows 95. It is provided so that you can adapt the scopes of work and inspection reports to suit your company's needs. The files on this diskette can only be used as word processed files on IBM compatible computers. Do not attempt to boot from this CD.

The scopes or work and inspection reports on this CD are in the form of Word 97/Windows 95 text files. More recent versions of Word, such as 98, should convert these documents into a useable format. Other word processing software (such as Ami Pro and WordPerfect) include a format conversion function. If you have word processing software other than Word, refer to your user's manual to determine how to convert these documents.

These files are provided so that you may modify and customize them as you would any other document in Word. However, if you experience technical problems with Word, consult your user's manual.

Document	Book Page	File Name
Block Foundations	25	Block Foundations
Block Foundations Inspection Report	29	Block Foundations IR
Brick Labor	31	Brick Labor
Brick Labor Inspection Report	35	Brick Labor IR
Cabinets	36	Cabinets
Cabinets Inspection Report	40	Cabinets IR
Cleaning	42	Cleaning
Cleaning (Rough) Inspection Report	46	Cleaning Rough IR
Cleaning (Final) Inspection Report	47	Cleaning Final IR
Clearing and Grading	49	Clearing-Grading
Clearing and Grading (Grading 1) Inspection Report	52	Clearing-Grading 1 IR
Clearing and Grading (Grading 2) Inspection Report	53	Clearing-Grading 2 IR
Clearing and Grading (Grading 3) Inspection Report	54	Clearing-Grading 3 IR
Deck and Porch Labor	55	Deck-Porch Labor
Deck and Porch Labor Inspection Report	59	Deck-Porch Labor IR
Drives and Walkways	60	Drives-Walkways
Drives and Walkways Inspection Report	64	Drives-Walkways IR
Drywall	65	Drywall
Drywall Inspection Report	70	Drywall IR
Electric	72	Electric
Electric (Rough) Inspection Report	76	Electric Rough IR
Electric (Final) Inspection Report	77	Electric Final IR
Fireplaces	78	Fireplaces
Fireplaces (Rough) Inspection Report	81	Fireplaces Rough IR
Fireplaces (Final) Inspection Report	82	Fireplaces Final IR
Floor Coverings	83	Floor Coverings
Floor Coverings Inspection Report	88	Floor Coverings IR
Floor Coverings (Carpet) Inspection Report	90	Floor Coverings Carpet IR
Footings	92	Footings
Footings Inspection Report	96	Footings IR
Framing Labor	97	Framing Labor
Framing Labor Inspection Report	104	Framing Labor IR
Garage Doors	107	Garage Doors
Garage Doors Inspection Report	110	Garage Doors IR
General Terms and Conditions	225	Gen Terms-Conditions
Guttering	111	Guttering

Document	Book Page	File Name
Guttering Inspection Report	114	Guttering IR
Heating and Air Conditioning	116	HVAC
Heating and Air Conditioning (Rough) Inspection Report	121	HVAC Rough IR
Heating and Air Conditioning (Final) Inspection Report	123	HVAC Final IR
Insulation	125	Insulation
Insulation Inspection Report	129	Insulation IR
Landscaping	131	Landscaping
Landscaping Inspection Report	135	Landscaping IR
Mirrors	137	Mirrors
Mirrors Inspection Report	140	Mirrors IR
Painting	141	Painting
Painting (Rough) Inspection Report	147	Painting Rough IR
Painting (Final) Inspection Report	149	Painting Final IR
Painting (Touch-Up) Inspection Report	151	Painting T-U IR
Plumbing	152	Plumbing
Plumbing (Slab) Inspection Report	157	Plumbing Slab IR
Plumbing (Rough) Inspection Report	158	Plumbing Rough IR
Plumbing (Final) Inspection Report	159	Plumbing Final IR
Poured Wall Foundations	161	Poured Walls
Poured Wall Foundations Inspection Report	165	Poured Walls IR
Roofing Labor	167	Roofing Labor
Roofing Labor Inspection Report	171	Roofing Labor IR
Shelving	172	Shelving
Shelving Inspection Report	175	Shelving IR
Shower Doors	176	Shower Doors
Shower Doors Inspection Report	179	Shower Doors IR
Shutters	180	Shutters
Shutters Inspection Report	183	Shutters IR
Siding and Cornice Labor	184	Siding-Cornice
Siding and Cornice Labor Inspection Report	188	Siding-Cornice IR
Slabs	189	Slabs
Slabs Inspection Report	194	Slabs IR
Stairs (Prebuilt)	195	Stairs
Stairs (Prebuilt) Inspection Report	198	Stairs IR
Supplier Purchase Order	229	Supplier PO
Terms and Conditions	224	Terms-Conditions
Trade Contractor Purchase Order	227	Contractor PO
Trim Labor	199	Trim Labor
Trim Labor Inspection Report	204	Trim Labor IR
Trim Labor (Lock-Out) Inspection Report	206	Trim Labor L-O IR
Termite Treatments	207	Termite Treatments
Termite Treatments Inspection Report	210	Termite Treatments IR
Vanity Tops	211	Vanity Tops
Vanity Tops Inspection Report	214	Vanity Tops IR
Waterproofing	215	Waterproofing
Waterproofing Inspection Report	218	Waterproofing IR
Window and Door Installation	219	Window-Door
Window and Door Installation Inspection Report	222	Window-Door IR